W9-ARA-101

Game Theory 101: The Complete Textbook

William Spaniel

Copyright William Spaniel, 2011-2014.

All rights reserved.

Acknowledgements

I thank Varsha Nair for her revisions as I compiled this book. I am also indebted to Kenny Oyama and Matt Whitten for their further suggestions. I originally learned game theory from Branislav Slantchev, John Duggan, and Mark Fey.

Please report possible errors to williamspaniel@gmail.com. I am grateful to those who have already given me feedback through this medium, and I encourage you to comment.

Table of Contents

Lesson 1.1: The Prisoner's Dilemma and Strict Dominance

At its core, game theory is the study of strategic interdependence—that is, situations where my actions affect both my welfare and your welfare and vice versa. Strategic interdependence is tricky, as actors need to anticipate, act, and react. Blissful ignorance will not cut it.

The prisoner's dilemma is the oldest and most studied model in game theory, and its solution concept is also the simplest. As such, we will start with it. Two thieves plan to rob an electronics store. As they approach the backdoor, the police arrest them for trespassing. The cops suspect that the pair planned to break in but lack the evidence to support such an accusation. They therefore require a confession to charge the suspects with the greater crime.

Having studied game theory in college, the interrogator throws them into the prisoner's dilemma. He individually sequesters both robbers and tells each of them the following:

We are currently charging you with trespassing, which implies a one month jail sentence. I know you were planning on robbing the store, but right now I cannot prove it—I need your testimony. In exchange for your cooperation, I will dismiss your trespassing charge, and your partner will be charged to the fullest extent of the law: a twelve month jail sentence.

I am offering your partner the same deal. If both of you confess, your individual testimony is no longer as valuable, and your jail sentence will be eight months each.

If both criminals are self-interested and only care about minimizing their jail time, should they take the interrogator's deal?

1.1.1: Solving the Prisoner's Dilemma

The story contains a lot of information. Luckily, we can condense everything we need to know into a simple matrix:

	Quiet	Confess
Quiet	-1, -1	-12, 0
Confess	0, -12	-8, -8

We will use this type of game matrix regularly, so it is important to understand how to interpret it. There are two players in this game. The first player's strategies (keep "quiet" and "confess") are in the rows, and the second player's strategies are in the columns. The first player's payoffs are listed first for each outcome, and the second player's are listed second. For example, if the first player keeps quiet and the second player confesses, then the game ends in the top right set of payoffs; the first player receives twelve months of jail time and the second player receives zero. Finally, as a matter of convention, we refer to the first player as a man and the second player as a woman; this will allow us to utilize pronouns like "he" and "she" instead of endlessly repeating "player 1" and "player 2."

Which strategy should each player choose? To see the answer, we must look at each move in isolation. Consider the game from player 1's perspective. Suppose he knew player 2 will keep quiet. How should he respond?

Let's focus on the important information in that context. Since player 1 only cares about his time in jail, we can block out player 2's payoffs with question marks:

	Quiet
Quiet	-1, ?
Confess	0, ?

Player 1 should confess. If he keeps quiet, he will spend one month in jail. But if he confesses, he walks away. Since he prefers less jail time to more jail time, confession produces his best outcome.

Note that player 2's payoffs are completely irrelevant to player 1's decision in this context—if he knows that she will keep quiet, then he only needs to look at his own payoffs to decide which strategy to pick. Thus, the question marks could be any number at all, and player 1's optimal decision given player 2's move will remain the same.

On the other hand, suppose player 1 knew that player 2 will confess. What should he do? Again, the answer is easier to see if we only look at the relevant information:

	Confess
Quiet	-12, ?
Confess	-8, ?

Confession wins a second time: confessing leads to eight months of jail time, whereas silence buys twelve. So player 1 would want to confess if player 2 confesses.

Putting these two pieces of information together, we reach an important conclusion—player 1 is better off confessing regardless of player 2's strategy! Thus, player 1 can effectively ignore whatever he thinks player 2 will do, since confessing gives him less jail time in either scenario.

Let's switch over to player 2's perspective. Suppose she knew that player 1 will keep quiet, even though we realize he should not. Here is her situation:

	Quiet	Confess
Quiet	?, -1	?, 0

As before, player 2 should confess, as she will shave a month off her jail sentence if she does so.

Finally, suppose she knew player 1 will confess. How should she respond?

	Quiet	Confess
Confess	?, -12	?, -8

Unsurprisingly, she should confess and spend four fewer months in jail.

Once more, player 2 prefers confessing regardless of what player 1 does. Thus, we have reached a solution: both players confess, and both players spend eight months in jail. The justice system has triumphed, thanks to the interrogator's savviness.

This outcome perplexes a lot of people new to the field of game theory. Compare the <quiet, quiet> outcome to the <confess, confess> outcome:

	Quiet	Confess
Quiet	-1, -1	?, ?
Confess	?, ?	-8, -8

Looking at the game matrix, people see that the <quiet, quiet> outcome leaves both players better off than the <confess, confess> outcome. They then wonder why the players cannot coordinate on keeping quiet. But as we just saw, promises to remain silent are unsustainable. Player 1 *wants* player 2 to keep quiet so when he confesses he walks away free. The same goes for player 2. As a result, the <quiet, quiet> outcome is inherently unstable. Ultimately, the players finish in the inferior (but sustainable) <confess, confess> outcome.

1.1.2: The Meaning of the Numbers and the Role of Game Theory

Although a large branch of game theory is devoted to the study of expected utility, we generally consider each player's payoffs as a ranking of his most preferred outcome to his least preferred outcome. In the prisoner's dilemma, we *assumed* that players only wanted to minimize their jail time. Game theory does *not* force players to have these preferences, as critics frequently claim. Instead, game theory analyzes what should happen given what players desire. So if players only want to minimize jail time, we could use the negative number of months spent in jail as their payoffs. This preserves their individual orderings over outcomes, as the most preferred outcome is worth 0, the least preferred outcome is -12, and everything else logically follows in between.

Interestingly, the cardinal values of the numbers are irrelevant to the outcome of the prisoner's dilemma. For example, suppose we changed the payoff matrix to this:

	Quiet	Confess
Quiet	3, 3	1, 4
Confess	4, 1	2, 2

Here, we have replaced the months of jail time with an ordering of most to least preferred outcomes, with 4 representing a player's most preferred outcome and 1 representing a player's least preferred outcome. In other words, player 1 would most like to reach the <confess, quiet> outcome, then the <quiet, quiet> outcome, then the <confess, confess> outcome, then the <quiet, confess> outcome.

Even with these changes, confess is still always better than keep quiet. To see this, suppose player 2 kept quiet:

	Quiet
Quiet	3, ?
Confess	4, ?

Player 1 should confess, since 4 beats 3.

Likewise, suppose player 2 confessed:

	Confess
Quiet	1, ?
Confess	2, ?

Then player 1 should still confess, as 2 beats 1.

The same is true for player 2. First, suppose player 1 kept quiet:

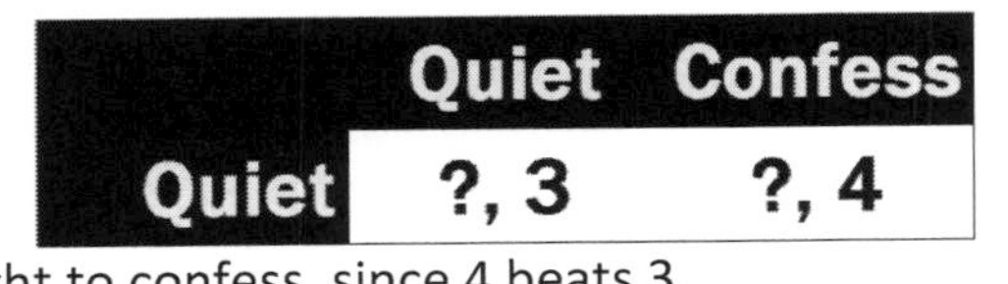

	Quiet	Confess
Quiet	?, 3	?, 4

Player 2 ought to confess, since 4 beats 3.

Alternatively, if player 1 confessed:

	Quiet	Confess
Confess	?, 1	?, 2

Player 2 should confess as well, as 2 is greater than 1. Thus, regardless of what the other player does, each player's best strategy is to confess.

To be clear, this preference ordering exclusively over time spent in jail is just one way the players may interpret the situation. Suppose you and a friend were actually arrested and the interrogator offered you a similar deal. The results here do not generally tell you what to do in that situation, unless you and your friend only cared about jail time. Perhaps your friendship is strong, and both of you value it more than avoiding jail time. Since confessing might destroy the friendship, you could prefer to keep quiet if your partner kept quiet, which changes the ranking of your outcomes. Your preferences here are perfectly rational. However, we do not yet have the tools to solve the corresponding game. We will reconsider these alternative sets of preferences in Lesson 1.3.

Indeed, the possibility of alternative preferences highlights game theory's role in making predictions about the world. In general, we take a three step approach:

1) Make assumptions.
2) Do some math.
3) Draw conclusions.

We do steps 1 and 3 everyday. However, absent rigorous logic, some conclusions we draw may not actually follow from our assumptions. Game theory—the math from step 2 that this book covers—provides a rigorous way of ensuring that that our conclusions follow directly from the assumptions. Thus, correct assumptions imply correct conclusions. But *incorrect* assumptions could lead to ridiculous claims. As such, we must be careful (and precise!) about the assumptions we make, and we should not be surprised if our conclusions change based on the assumptions we make.

Nevertheless, for the given payoffs in the prisoner's dilemma, we have seen an example of *strict dominance*. We say that a strategy x strictly dominates strategy y for a player if strategy x provides a greater payoff for that player than strategy y regardless of what the other players do. In this example, confessing strictly dominated keeping quiet for both players. Unsurprisingly, players never optimally select strictly dominated strategies—by definition, a better option *always* exists regardless of what the other players do.

1.1.3: Applications of the Prisoner's Dilemma

The prisoner's dilemma has a number of applications. Let's use the game to explore optimal strategies in a number of different contexts.

First, consider two states considering whether to go to war. The military technology available to these countries gives the side that strikes first a large advantage in the fighting. In fact, the first-strike benefit is so great that each country would prefer attacking the other state even if its rival plays a peaceful strategy. However, because war destroys property and kills people, both prefer remaining at peace to simultaneously declaring war.

Using these preferences, we can draw up the following matrix:

	Defend	Attack
Defend	3, 3	1, 4
Attack	4, 1	2, 2

From this, we can see that the states most prefer attacking while the other one plays defensively. (This is due to the first-strike advantage.) Their next best outcome is to maintain the peace through mutual defensive strategies. After that, they prefer declaring war simultaneously. Each state's worst outcome is to choose defense while the other side acts as the aggressor.

We do not need to solve this game—we already have! This is the same game from the previous section, except we have exchanged the labels "quiet" with "defend" and "confess" with "attack." Thus, we know that both states attack in this situation even though they both prefer the <defend, defend> outcome. The first-strike advantages trap the states in a prisoner's dilemma that leads to war.

A similar problem exists with arms races. Imagine states must simultaneously choose whether to develop a new military technology. Constructing weapons is expensive but provides greater security against rival states. We can draw up another matrix for this scenario:

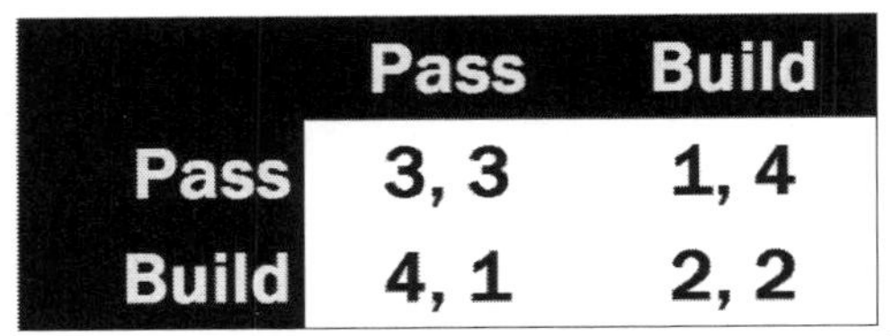

	Pass	Build
Pass	3, 3	1, 4
Build	4, 1	2, 2

Here, the states most prefer building while the other state passes. Following that, they prefer the <pass, pass> outcome to the <build, build> outcome; the states maintain the same relative military strength in both these outcomes, but they do not waste money on weaponry if they both pass. The worst possible outcome is for the other side to build while the original side passes. Again, we already know the solution to this game. Both sides engage in the arms race and build.

Now consider international trade. Many countries place tariffs (a tax) on imported goods to protect domestic industries even though this leads to higher prices overall.

We can use the prisoner's dilemma to explain this phenomenon. A country can levy a tariff against another country's goods or opt for no taxes. The best outcome for a country is to tax imports while not having the other country tax its exports. This allows the domestic industries to have an advantage at home and be competitive abroad, and the country also earns revenue from the tax itself. Free trade is the next best outcome, as it allows the lowest prices for each country's consumers. Mutual tariffs is the next best outcome, as they give each country an advantage at home but a disadvantage abroad; ultimately, this leads to higher prices than the free trade outcome. The worst possible outcome is to levy no taxes while the other country enforces a tariff, as domestic industries stand no chance against foreign rivals.

Let's toss that information into another matrix:

	No Tax	Tax
No Tax	3, 3	1, 4
Tax	4, 1	2, 2

We know this is a prisoner's dilemma and both sides will tariff each other's goods: taxing strictly dominates not taxing in this setup.

Finally, consider two rival firms considering whether to advertise their products. Would the firms ever want the government to pass a law forbidding advertisement? Surprisingly, if advertising campaigns only persuade a consumer to buy a certain brand of product rather than the product in general, the answer is yes. If one side places ads and the other does not, the firm with the advertising campaign cuts into the other's share of the market. If they both advertise, the ads cancel each other out, but they still have to pay for the campaigns.

If we look at the corresponding matrix, we see another classic example of the prisoner's dilemma:

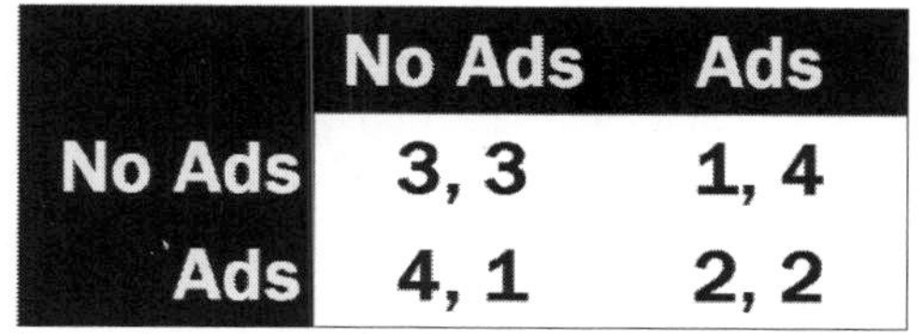

	No Ads	Ads
No Ads	3, 3	1, 4
Ads	4, 1	2, 2

Thus, both sides advertise to preempt the other side's campaign. The ads ultimately cancel each other out, and the firms end the game in a worse position than had they both not placed ads.

The Public Health Cigarette Smoking Act is a noteworthy application of the advertising game. In 1970, Richard Nixon signed the law, which removed cigarette ads from television. Tobacco companies actually *benefited* from this law in a perverse way—the law forced them to cooperating with each other. In terms of the game matrix, the law pushed them from the <2, 2> payoff to the mutually preferable <3, 3> payoff. The law simultaneously satisfied politicians, as it made targeting children more difficult for all tobacco companies.

These examples illustrate game theory's ability to draw parallels between seemingly dissimilar situations. We have seen models of prisoner confession, wars, arms races, taxation, and advertisements. Despite the range of examples, each had an underlying prisoner's dilemma mechanism. In this manner, game theory allows us to unify a wide-range of life decisions under a single, unified framework.

1.1.4: Deadlock

The 2012 Summer Olympics badminton tournament provides an interesting case study of strategic manipulation. The tournament featured round-robin group play with a cut to a single-elimination quarterfinals bracket. Officials determined the seeding for the quarterfinals by the win/loss records during the round-robin matches.

In the morning matches of the final day of round-robin play, the second-best team in the world lost. While their previous victories still ensured that the team would reach the quarterfinals, their defeat pushed them into the lower half of the seeding. This had an interesting impact on the afternoon matches. Teams who had already clinched a quarterfinal spot now had incentive to *lose* their remaining games. After all, a higher seeding meant a greater likelihood of facing the world's second-best team earlier in the elimination rounds. Matches turned into contests to see who could lose most efficiently!

To untangle the twisted logic at work here, consider the following game. Two players have to choose whether to *try* or *fail*. The quality of any corresponding outcome is diagrammed in the game matrix below:

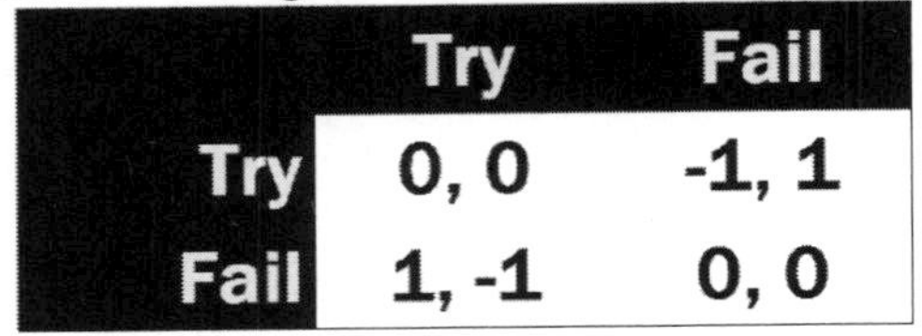

	Try	Fail
Try	0, 0	-1, 1
Fail	1, -1	0, 0

Ordinarily, we would expect trying to be a good thing and failing to be a bad thing. The reverse is true here. Each team most prefers failing while the other team tries; this ensures the team in question will lose, drop into the lower part of the quarterfinals bracket, and thus avoid the world's second best team. The worst outcome for a team is for that team to try while the other team fails; this ensures that the original team wins the match but then must face a harder path through the single elimination bracket. If both try or both fail, then neither has an inherent strategic advantage.

Like the prisoner's dilemma, we can solve this game with strict dominance alone. Here, fail dominates try for both parties. We can verify this using the same process as before. First, suppose player 2 chooses tries:

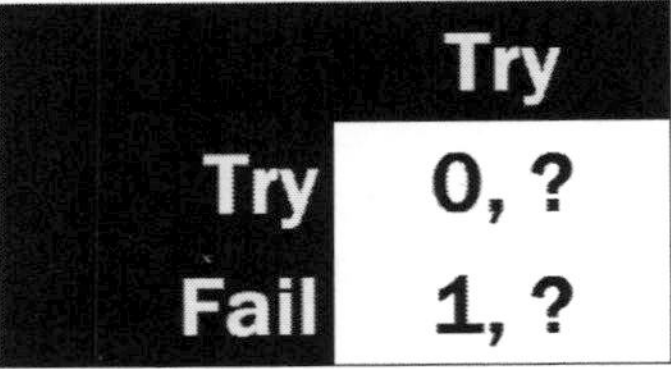

	Try
Try	0, ?
Fail	1, ?

If player 1 tries, he earns 0; if he fails, he earns 1. Since 1 beats 0, he should fail in this situation.

Now consider player 1's response to player 2 failing:

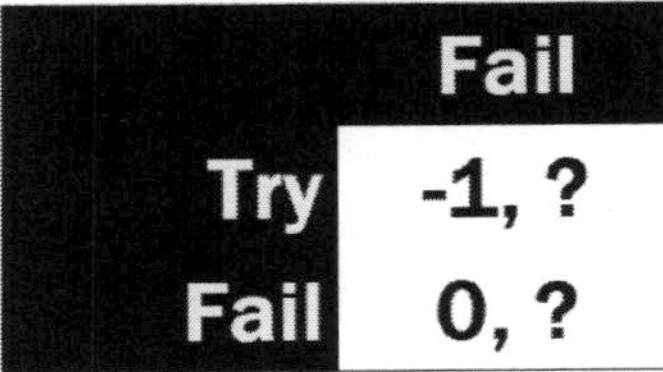

	Fail
Try	-1, ?
Fail	0, ?

Again, fail triumphs: failing nets him 0 while trying earns him -1. Because failing is better than trying for player 1 regardless of player 2's strategy, fail strictly dominates try for him.

The same story holds for player 2. Consider her response to player 1 trying:

	Try	Fail
Try	?, 0	?, 1

If player 2 tries, she earns 0; if she fails, she earns 1. Thus, she ought to fail in this situation.

Now suppose player 1 failed instead:

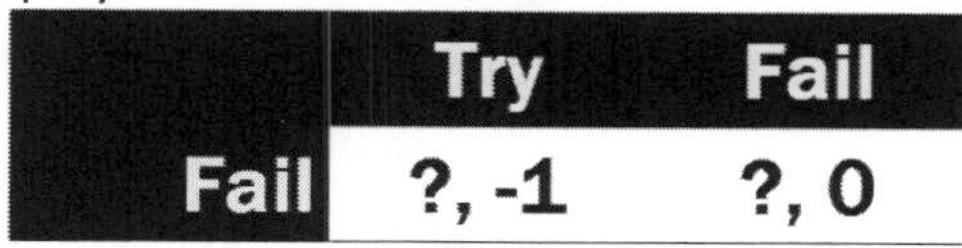

	Try	Fail
Fail	?, -1	?, 0

Player 2 still ought to fail: -1 is less than 0. As a result, fail strictly dominates try for her as well. In turn, we should expect both of them to fail. Despite the absurdity of the outcome, the perverse incentives of the tournament structure make intentionally failing a sensible strategy!

This badminton example is a slight modification of a generic game called *deadlock*. It gets its name because the players cannot improve the quality of their outcomes unless the opponent chooses his or her strategy incorrectly. Here are the generic game's payoffs:

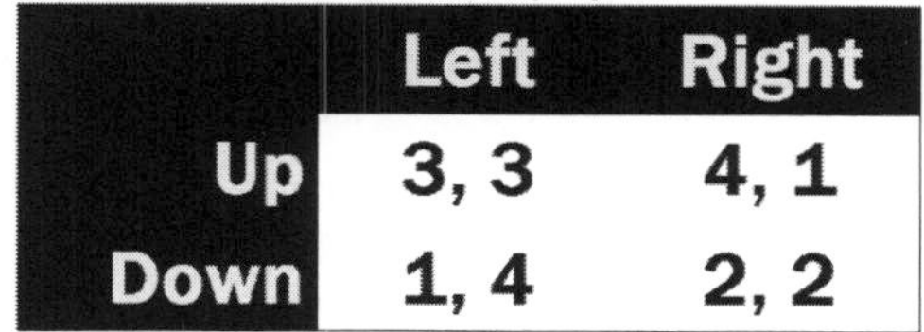

	Left	Right
Up	3, 3	4, 1
Down	1, 4	2, 2

Again, we can solve this game using strict dominance. Specifically, up strictly dominates down and left strictly dominates right. Let's verify this, starting with player 1's strategy:

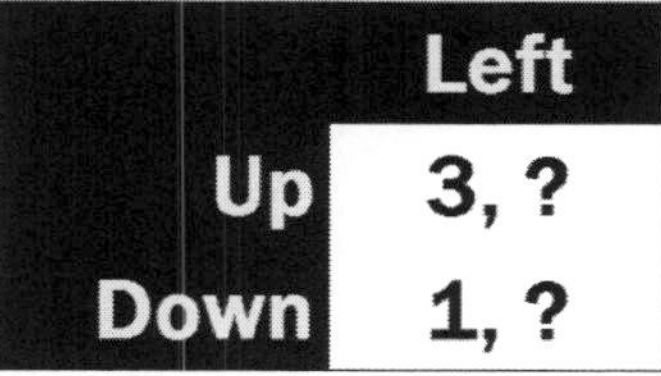

	Left
Up	3, ?
Down	1, ?

We see that up is better than down, as 3 beats 1.

Repeating this for right, we focus on the following:

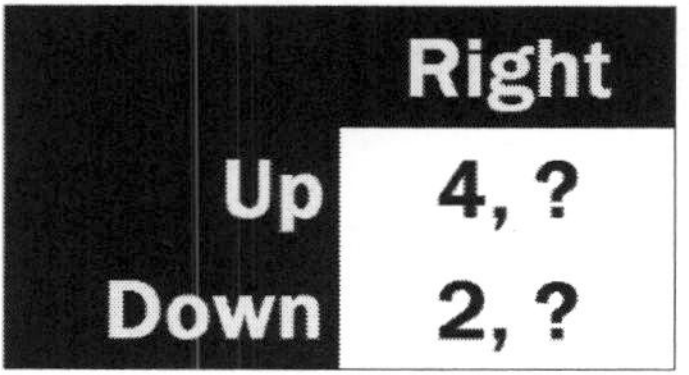

	Right
Up	4, ?
Down	2, ?

Once more, up is better than down, since 4 beats 2. So up is a better strategy than down regardless of what player 2 does.

Switching gears, suppose player 1 selected up. Then player 2 can focus on the following contingency:

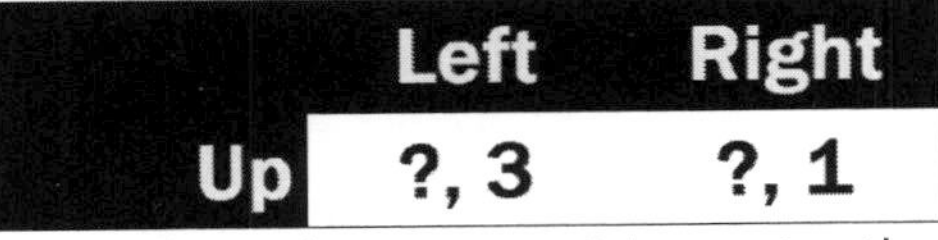

	Left	Right
Up	?, 3	?, 1

Left is better than right in this case, as 3 is greater than 1.

Repeating this process a final time, player 2 now assumes player 1 will play down:

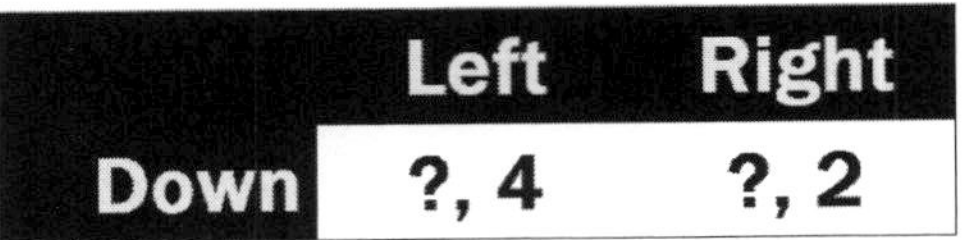

	Left	Right
Down	?, 4	?, 2

Left is still better than right, as 4 is greater than 2. Since left always beats right regardless of what player 1 does, left strictly dominates right, and therefore player 2 will play left. Thus, the outcome is <up, left>.

Thus, both players are locked into their strictly dominant strategy and will never achieve their best outcome unless the other makes a mistake. However, unlike in the prisoner's dilemma, no alternative outcome exists that is simultaneously better for both players than the <up, left> solution. As such, deadlock may be more intuitive, but it also tends to be substantively less interesting.

1.1.5: Strict Dominance in Asymmetric Games

We can use strict dominance on games even when they are not symmetric like the prisoner's dilemma or deadlock. For example, consider the arms race from earlier. Suppose that player 2 maintains her same payoffs. That is, she most prefers arming while her opponent passes and least prefers the opposite outcome. Meanwhile, she prefers neither side arming to both arming, as the balance of power remains the same but she saves on the costs of weapons. On the other hand, suppose that player 1 is a pacifist. He simply receives -1 for each party that builds weapons.

With that, we can construct the following payoff matrix:

	Pass	Build
Pass	0, 3	-1, 4
Build	-1, 1	-2, 2

Unlike before, each player has a distinct set of payoffs. But if we run through the same process as before, we will see that <pass, build> is the only reasonable solution.

Let's begin with player 1's choices. Suppose player 2 moved passes. How should player 1 respond?

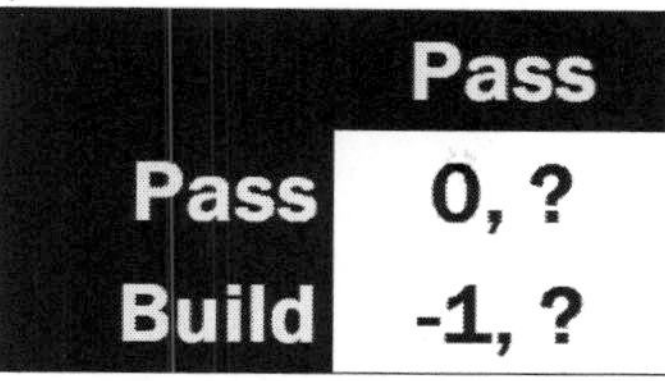

	Pass
Pass	0, ?
Build	-1, ?

Recall that player 1 wants to minimize the total number of weapons. If he passes while player 2 passes, he achieves his best possible outcome. If he builds, he receives a -1. As such, he would want to pass in this situation.

Now suppose player 2 chose builds. Again, we need to find how player 1 should optimally respond:

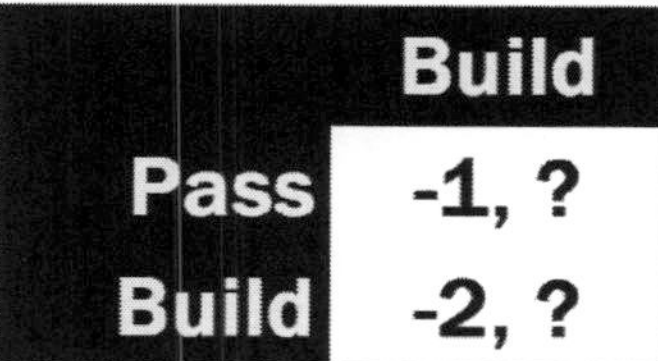

	Build
Pass	-1, ?
Build	-2, ?

This time, player 1 cannot reach his best possible outcome. He can, however, minimize his losses by passing instead of building. Consequently, he would pass in this situation as well.

Combining the last two inferences together, we know player 1's optimal strategy: he will pass regardless of how player 2 behaves.

That leaves us to solve for player 2's strategy. Let's start she should respond to pass:

	Pass	Build
Pass	?, 3	?, 4

Player 2 can achieve her best possible outcome here by building, since she can exploit the shift in power. Since 4 beats 3, she will build in this situation.

Now suppose player 1 builds instead:

	Pass	Build
Build	?, 1	?, 2

Although player 2 can no longer reach her favorite outcome, she can at least keep pace with player 1's power by building here. As such, she would build if she knew that player 1 would build.

Despite the game's asymmetry, the game still has a solution in dominant strategies: <pass, build>. Player 2 achieves her best outcome, while player 1 must settle for a moderate result since he cannot stop her from arming.

Conclusion

Overall, strict dominance is a powerful tool in game theory. But while the concept is simple, applying it can be difficult. Even in matrix form, a game still has a lot of information. To successfully find dominated strategies, we must focus on one player's payoffs at a time. Above, we used question marks to isolate the relevant payoffs. When searching for strictly dominated strategies on your own, mentally block out the irrelevant payoffs and strategies in a similar manner.

Takeaway Points

1) Game theory is a mathematical method to ensure that assumptions imply conclusions.
2) Payoffs in a game matrix represent a player's preferences according to the assumptions.
3) Strategy x *strictly dominates* strategy y if it produces a higher payoff than y regardless of what all other players do.
4) Playing a strictly dominated strategy is irrational—another strategy always yields a better outcome.

Lesson 1.2: Iterated Elimination of Strictly Dominated Strategies

Analyzing how companies optimally act in a world of imperfect competition is one of game theory's greatest strengths. When the number of companies in the world is arbitrarily large, a single company cannot manipulate the market on its own. But if only two rival companies exist, their individual business decisions have a great impact on the others.

Let's see this in action. Suppose a small town has only two dance clubs, called ONE and TWO. Both are deciding whether to host a salsa night or disco night this Friday. Club ONE has a strategic advantage over club TWO: ONE is located at the center of the town, while TWO is a few miles away. Thus, if TWO runs the same theme as ONE, nobody will show up to TWO.

There are three types of customers. 60 hardcore salsa fans will only go to a club if salsa is being offered. 20 people are hardcore disco fans and will only go to a club if disco is being offered. A final 20 people prefer going to a disco theme but will attend a salsa night if that is the only option.

If the businesses want to maximize the number of customers on Friday, the payoff matrix looks like this:

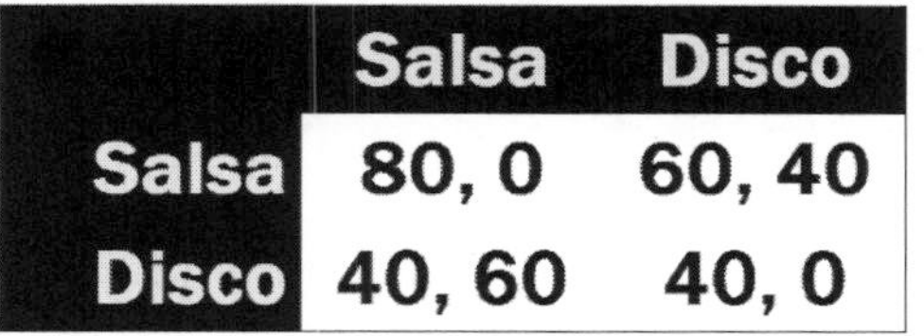

	Salsa	Disco
Salsa	80, 0	60, 40
Disco	40, 60	40, 0

Note that TWO appears to be in a dilemma. If it offers the same theme as ONE, nobody will show up. Thus, TWO's ultimate goal must be to successfully predict ONE's theme and then adopt the alternative. Consequently, TWO has no strictly dominant strategy.

Should TWO be resigned to just guessing which strategy to play? Absolutely not. If TWO unravels ONE's incentives, TWO will know exactly which theme to offer.

To see how, consider which strategy ONE should play. First, suppose ONE anticipates TWO will choose salsa. How should ONE respond?

	Salsa
Salsa	80, ?
Disco	40, ?

If ONE picks salsa, it brings in 80 customers. If it selects disco, it receives only 40. Since 80 is greater than 40, ONE should play salsa in response to TWO choosing salsa.

Now suppose ONE believes TWO will plan a disco night:

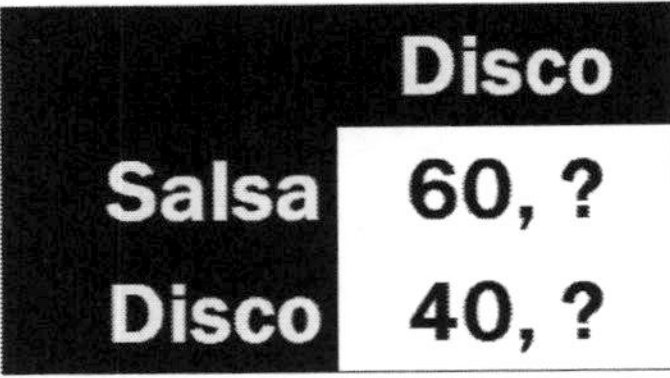

	Disco
Salsa	60, ?
Disco	40, ?

Once again, salsa is better; salsa brings in 60 customers, whereas disco generates only 40.

Stringing these two pieces of information together, we know salsa strictly dominates disco for ONE—regardless of TWO's selection, ONE is always better off choosing salsa. Therefore, ONE must optimally have a salsa night.

With that in mind, consider how TWO should reason. Putting itself into ONE's shoes, TWO realizes that ONE will have a salsa night. Thus, TWO's strategic dilemma boils down to the following choice:

	Salsa	Disco
Salsa	?, 0	?, 40

If TWO also plans a salsa night, everyone attends ONE due to ONE's superior location, and TWO ends up with no customers for the evening. Alternatively, if TWO opts for a disco night, all 40 of the disco fans show up. Consequently, TWO optimally plays disco. Therefore, the only rational outcome of the game is <salsa, disco>.

While the logic of the club game appears straightforward, we can extend it to more complex situations. Consider the following game, where the numbers represent the dollars won (or lost) by each player for the particular outcomes:

	Left	Center	Right
Up	13, 3	1, 4	7, 3
Middle	4, 1	3, 3	6, 2
Down	-1, 9	2, 8	8, -1

If we suppose the players only want to maximize the number of dollars they win, this game presents a new challenge. With the prisoner's dilemma, we knew exactly what each player would do—confessing was always better than remaining silent regardless of the other player's strategy. In the club game, ONE had a single strictly dominant strategy, and we could find TWO's optimal strategy based off of that knowledge.

This game is not as simple. For example, player 1 is in a quandary. Suppose he knew player 2 would go left:

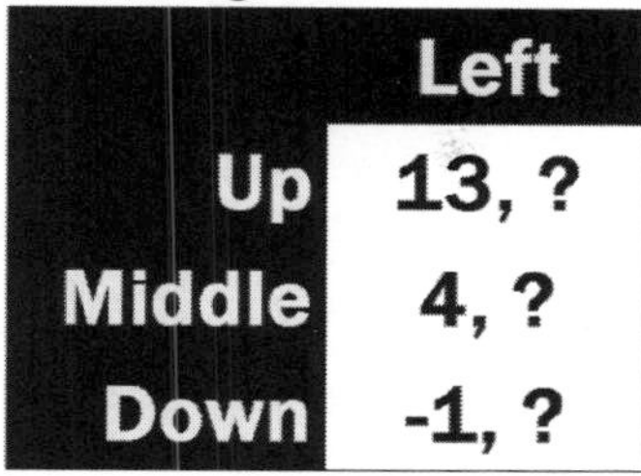

	Left
Up	13, ?
Middle	4, ?
Down	-1, ?

Player 1 would then want play up, as the 13 he earns is better than the 4 he earns for middle and -1 he earns for down. But this is not the case if player 2 chose center:

	Center
Up	1, ?
Middle	3, ?
Down	2, ?

Now player 1 ought to play middle, as the 3 is better than the 1 for going up and the 2 for going down.

And to make things more complicated, look what happens if player 1 knew player 2 would choose right:

	Right
Up	7, ?
Middle	6, ?
Down	8, ?

In this case, player 1 should go down and earn 8.

So in three different cases, player 1 should do three different things. How should player 1 decide which to choose?

1.2.1: Using Iterated Elimination of Strictly Dominated Strategies

In the previous lesson, we discussed why players ought to never play strictly dominated strategies. If player are intelligent, they should infer how others will *not* act and tailor their strategies accordingly.

Let's generalize this thought process. First, take another look at the prisoner's dilemma:

	Quiet	Confess
Quiet	-1, -1	-12, 0
Confess	0, -12	-8, -8

Recall that confessing strictly dominates keeping quiet for player 1. That is, if player 2 were to keep quiet, player 1 would want to confess (0 is better than -1); and if player 2 were to confess, player 1 would want to confess (-8 is better than -12). So player 1 would never keep quiet.

Now switch to player 2's perspective. She knows that player 1 is intelligent enough to see that confessing strictly dominates keeping quiet. Therefore, she infers that he will never keep quiet. Given that, for all intents and purposes, player 2 can ignore keep quiet as a strategy for player 1—he will never play it. As such, from her perspective, she can reduce the game to the following:

	Quiet	Confess
Confess	0, -12	-8, -8

At this point, player 2 should confess and earn -8 rather than keep quiet and earn -12. Note that she never even had to observe that confessing strictly dominated keeping quiet for her as well; instead, she merely reduced the game by removing player 1's implausible choice. And since the only strategy that remaining for player 1 is confess, player 2 can condition her response purely based off of that information.

This process is known as *iterated elimination of strictly dominated strategies* (IESDS). We began by noting that confessing strictly dominated keeping quiet for player 1, so we erased keep quiet as a strategy for him. We then pretended that the remaining game was all that mattered and found that confess strictly dominated keep quiet for player 2.

IESDS takes some complicated games and reduces them into simpler games. We can see this in action using the 3x3 game from earlier:

	Left	Center	Right
Up	13, 3	1, 4	7, 3
Middle	4, 1	3, 3	6, 2
Down	-1, 9	2, 8	8, -1

Let's isolate player 2's choice between center and right:

	Center	Right
Up	?, 4	?, 3
Middle	?, 3	?, 2
Down	?, 8	?, -1

Should player 2 select play right? She should not—center strictly dominates right. To see this, note that if player 1 chooses up, then center beats right, as 4 is greater than 3. Likewise, if he plays middle, center's 3 is superior to right's 2. Finally, if player 1 plays down, center defeats right 8 to -1. Consequently, regardless of which strategy player 1 chooses, center is better than right for player 2.

Now look at the game from player 1's perspective. He knows that player 2 will never play right, so he only needs to consider the following game:

	Left	Center
Up	13, 3	1, 4
Middle	4, 1	3, 3
Down	-1, 9	2, 8

Consider player 1's decision between middle and down:

	Left	Center
Middle	4, ?	3, ?
Down	-1, ?	2, ?

Down is no longer a sensible choice—middle strictly dominates it, as 4 beats -1 and 3 beats 2. Essentially, down was only useful as a contingency plan if player 2 played right. But since right is not a sensible choice for player 2, she will not play it. And since player 1 knows that she is sensible and will not play right, he has no reason to play down.

Let's flip back to player 2's perspective. She knows that right is a lousy choice for her. She also knows player 1 knows that she knows this. As a result, player 1 infers that down is a lousy choice for him. Working through that logic, player 2 knows he will not play down, and therefore she can reduce the original game to the following:

	Left	Center
Up	13, 3	1, 4
Middle	4, 1	3, 3

Center now strictly dominates left for her:

	Left	Center
Up	?, 3	?, 4
Middle	?, 1	?, 3

We see that if player 1 plays up, then center's 4 beats left's 3. And likewise, if player 1 selects middle, center is again better than left because 3 beats 1. So player 2 would never play left.

After working through another level of the he-knows-that-she-knows-that-he-knows logic, player 1 can conclude he is looking at the following reduced game:

	Center
Up	1, 4
Middle	3, 3

Isolating player 1's payoffs makes it clear that he will play middle since 3 beats 1:

	Center
Up	1, ?
Middle	3, ?

Therefore, we can conclude that player 1 will play middle and player 2 will play center.

1.2.2: Duopolistic Competition

When a single firm controls all of the production of a single good, we call it a monopoly. When exactly two competing firms control production an identical good, we call this economic environment a *duopoly*. Each firm's production affects the other's ability to profit. Thus, we can treat these two firms as players in a game of profit maximizing—a game that IESDS can solve.

Suppose Firm 1 and Firm 2 must spend $1 to produce a unit of a good. Consumer demand determines the price of the good; the larger the quantity available, the lower price. Let P be the consumer's market price of the good and Q be the total quantity of units the two firms produce. Here, if the firms collectively produce six or fewer units, the price demand function is as follows:

P = $12 – ($2)(Q)

If the firms collectively produce more than six units, then the market price is $0.

Note that the quantity produced is a function of both firm's strategic decisions. A single firm cannot control the other firm's production quantity, which in turn means it cannot unilaterally determine ultimate market price. As such, we might wonder if the firms have an optimal production strategy.

To begin, each firm has only six plausible production choices: zero, one, two, three, four, or five. Producing six more units is not sensible; each unit costs $1 to produce, but flooding the market with more than five units ensures the ultimate price will be less than $1, guaranteeing a net loss.

We can use this as a starting point to construct the payoff matrix. Since each firm has six remaining strategies, we are looking at a daunting 6x6 game:

	Zero	One	Two	Three	Four	Five
Zero	0, 0	0, 9	0, 14	0, 15	0, 12	0, 5
One	9, 0	7, 7	5, 10	3, 9	1, 4	-1, -5
Two	14, 0	10, 5	6, 6	2, 3	-2, -4	-2, -5
Three	15, 0	9, 3	3, 2	-3, -3	-3, -4	-3, -5
Four	12, 0	4, 1	-4, -2	-4, -3	-4, -4	-4, -5
Five	5, 0	-5, -1	-5, -2	-5, -3	-5, -4	-5, -5

To see how we could derive these payoffs ourselves, let's run through a few of the outcomes. To start, consider the outcome where Firm 1 produces three units and Firm 2 produces zero. Firm 2 obviously makes no profit, as it neither creates nor sells any units. Meanwhile, it costs Firm 1 $3 to produce three units. It then sells those units at the price that the market demand dictates:

$Q = 3 + 0$
$Q = 3$
$P = \$12 – (\$2)(Q)$
$P = \$12 – (\$2)(3)$
$P = \$12 – \$6$
$P = \$6$

So Firm 1 sells three units for $6 each, or $18. After subtracting the $3 in production costs, Firm 1 achieves a total profit of $15. Thus, in the payoff matrix, Firm 1 earns 15 and Firm 2 earns 0.

Now suppose both Firm 1 and Firm 2 produce two units. Both pay $2 in production. They then sell the goods at the market price:

$Q = 2 + 2$

Q = 4
P = \$12 – (\$2)(Q)
P = \$12 – (\$2)(4)
P = \$12 – \$8
P = \$4

So each firm sells two units for \$4, for \$8 in gross revenue. After subtracting the \$2 in production costs, the firms both take home \$6. Hence, in the payoff matrix, both players receive an expected utility of 6 if they both produce two units.

For the final example, suppose Firm 1 produced three units and Firm 2 produced four. This time, the firms have collectively produced more than five units. In that case, the market price is \$0—the firms have flooded the public with more goods than people are willing to pay for. Consequently, the firms eat their production costs and receive no revenue. In the payoff matrix, Firm 1 receives -3 and Firm 2 produces -4. Both firms would have been better off had not they not produced any units at all.

With a little patience, this game is easy to solve. First, compare Firm 1's payoffs for producing three units to its payoffs for producing five units:

	Zero	One	Two	Three	Four	Five
Three	15, ?	9, ?	3, ?	-3, ?	-3, ?	-3, ?
Five	5, ?	-5, ?	-5, ?	-5, ?	-5, ?	-5, ?

Regardless of Firm 2's production level, Firm 1 is always better off producing three units. If Firm 2 produces zero, three beats five 15 to 5. Against one, three remains optimal by the margin of 9 to -5. Against two, the margin is closer, but three still wins 3 to -5. Finally, if Firm 2 produces three, four, or five, Firm 1 is better off producing three units and earning -3 than producing five and earning -5.

Since the firms have symmetrical production capacities and sell identical products, three also strictly dominates five for Firm 2:

	Three	Five
Zero	?, 15	?, 5
One	?, 9	?, -5
Two	?, 3	?, -5
Three	?, -3	?, -5
Four	?, -3	?, -5
Five	?, -3	?, -5

The margins are exactly the same as in the previous case. Thus, we can eliminate five as a strategy for both firms, leaving the following remaining game:

	Zero	One	Two	Three	Four
Zero	0, 0	0, 9	0, 14	0, 15	0, 12
One	9, 0	7, 7	5, 10	3, 9	1, 4
Two	14, 0	10, 5	6, 6	2, 3	-2, -4
Three	15, 0	9, 3	3, 2	-3, -3	-3, -4
Four	12, 0	4, 1	-4, -2	-4, -3	-4, -4

Producing three also strictly dominates producing four for Firm 1:

	Zero	One	Two	Three	Four
Three	15, ?	9, ?	3, ?	-3, ?	-3, ?
Four	12, ?	4, ?	-4, ?	-4, ?	-4, ?

If Firm 2 produces zero, three beats four 15 to 12. Against one, three wins 9 to 4. Versus two, three reigns supreme 3 to -4. And if Firm 2 produces three or four, three trumps four -3 to -4.

Again, the strict dominance is symmetrical for Firm 2:

	Three	Four
Zero	?, 15	?, 12
One	?, 9	?, 4
Two	?, 3	?, -4
Three	?, -3	?, -4
Four	?, -3	?, -4

As a result, we can remove four from the game for both players. After doing so, we are left with the following reduced game:

	Zero	One	Two	Three
Zero	0, 0	0, 9	0, 14	0, 15
One	9, 0	7, 7	5, 10	3, 9
Two	14, 0	10, 5	6, 6	2, 3
Three	15, 0	9, 3	3, 2	-3, -3

Notice zero's relationship with one:

	Zero	One	Two	Three
Zero	0, ?	0, ?	0, ?	0, ?
One	9, ?	7, ?	5, ?	3, ?

If Firm 1 produces zero, it guarantees no profit at all. On the other hand, Firm 1 could produce one unit and guarantee some profit. Thus, producing nothing has become an unreasonable strategy for Firm 1.

Of course, zero is an equally unreasonable strategy for Firm 2 for the same reason:

	Zero	One
Zero	?, 0	?, 9
One	?, 0	?, 7
Two	?, 0	?, 5
Three	?, 0	?, 3

After we remove zero from the matrix, each firm is down to only three strategies:

	One	Two	Three
One	7, 7	5, 10	3, 9
Two	10, 5	6, 6	2, 3
Three	9, 3	3, 2	-3, -3

Now two strictly dominates three:

	One	Two	Three
Two	10, ?	6, ?	2, ?
Three	9, ?	3, ?	-3, ?

If Firm 2 produces one, two beats three 10 to 9. Versus two, two defeats three 6 to 3. Lastly, against three, two wins once more, 2 to -3.

As always, the same is true for Firm 2:

	Two	Three
One	?, 10	?, 9
Two	?, 6	?, 3
Three	?, 2	?, -3

If we remove these strictly dominated strategies, we are down to a simple 2x2 game:

	One	Two
One	7, 7	5, 10
Two	10, 5	6, 6

This game is just like a prisoner's dilemma from last lesson. Two strictly dominates one for both players. Here is the comparison for Firm 1:

	One	Two
One	7, ?	5, ?
Two	10, ?	6, ?

If Firm 2 produces one, Firm 2 prefers producing two and earning 10 rather than producing one and earning 7. Likewise, if Firm 1 produces two, two is still better for Firm 2 by a 6 to 5 margin.

The same holds for Firm 2:

	One	Two
One	?, 7	?, 10
Two	?, 5	?, 6

Thus, two strictly dominates one for both firms. Despite 36 outcomes in the original game, iterated elimination of strictly dominated strategies yields a single solution. Both firms produce two units. With four total units on the market, they sell for $4 each. The firms make $8 in gross revenue but must pay $2 in production costs for a net profit of $6 apiece.

Note that the <one, one> outcome is better for both firms than the <two, two> outcome, as they both receive $7 in net profits:

	One	Two
One	7, 7	5, 10
Two	10, 5	6, 6

However, the <one, one> outcome is analogous to the <quiet, quiet> outcome in the prisoner's dilemma. Although the firms would like to collude to reduce production quantities and in turn artificially inflate market prices, neither firm can credibly commit to that course of action. After all, if one firm reduces its quantity produced, market prices go up, and it becomes more tempting for the other firm to break the agreement. Here, if one side breaks the agreement on the <one, one> outcome, its profits shoot up to $10. Only the <two, two> outcome is inherently stable.

1.2.3: Does Order Matter?

Suppose we had a game that started with two strictly dominated strategies. A natural question is whether we will end up with a different answer depending on which one we eliminate first.

In fact, our first choice is irrelevant. The reason for this is a little complicated, so let's instead look at a couple of examples. Earlier in this section, we solved the prisoner's dilemma by eliminating player 1's keep quiet strategy first. Based off of the remaining game, we eliminated keep quiet for player 2. Thus, both players optimally confessed.

However, since we know confessing strictly dominates keeping quiet for player 2 in the original game, we could have started by removing her keep quiet strategy first. That would have left us with this game:

	Confess
Quiet	-12, 0
Confess	-8, -8

Unsurprisingly, confessing strictly dominates keeping quiet in this reduced game for player 1:

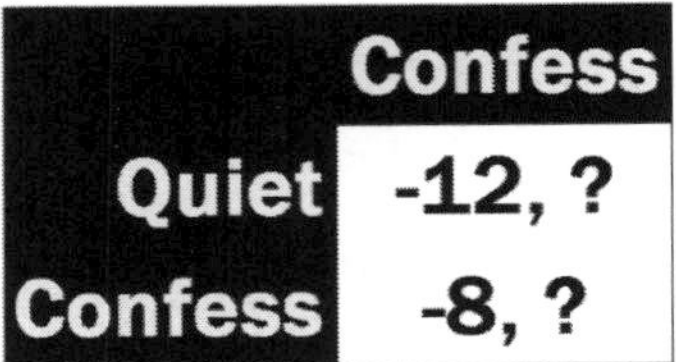

	Confess
Quiet	-12, ?
Confess	-8, ?

Since -8 beats -12, player 1 should confess. Thus, both players confess.

The prisoner's dilemma might seem like a trivial example given that the players face a symmetrical situation. Is order still irrelevant in asymmetric cases? Yes. For example, consider this game:

	Left	Right
Up	-1, 1	4, 2
Middle	0, 2	3, 3
Down	-2, -2	2, -1

A couple different IESDS paths can solve this game. First, note that middle strictly dominates down for player 1:

	Left	Right
Middle	0, ?	3, ?
Down	-2, ?	2, ?

That is, beats -2 and 3 beats 2. So player 1 would never want to play down, which reduces the original game to the following:

	Left	Right
Up	-1, 1	4, 2
Middle	0, 2	3, 3

Notice that right strictly dominates left for player 2:

	Left	Right
Up	?, 1	?, 2
Middle	?, 2	?, 3

This is because 2 beats 1 and 3 beats 2. Thus, we can eliminate left from the game:

	Right
Up	4, 2
Middle	3, 3

And now the game boils down to what is best for player 1:

	Right
Up	4, ?
Middle	3, ?

Since 4 is greater than 3, player 1 will play up. Thus, the solution is <up, right>.

Let's go back to the original game:

	Left	Right
Up	-1, 1	4, 2
Middle	0, 2	3, 3
Down	-2, -2	2, -1

While middle strictly dominates down, note that right also strictly dominates left:

	Left	Right
Up	?, 1	?, 2
Middle	?, 2	?, 3
Down	?, -2	?, -1

That is, 2 beats 1, 3 beats 2, and -1 is greater than -2. So instead of eliminating down to start, we could have eliminated left instead. Had we done so, the reduced game would have looked like this:

	Right
Up	4, 2
Middle	3, 3
Down	2, -1

From here, it is only a matter of selecting player 1's greatest payoff:

	Right
Up	4, ?
Middle	3, ?
Down	2, ?

Since 4 is greater than 3 or 2, player 1 plays up. Therefore, we end up at the <up, right> outcome regardless of which path we take. This holds for *any* game when we use iterated elimination of strictly dominated strategies. Consequently, when you are solving complex games and you find a strictly dominated strategy, *eliminate it immediately*. Although there may be more strategies you could eliminate in the first step, these strategies will still be strictly dominated in the next step. It will also be easier to find them, as there is less information to consider in the remaining game.

1.2.4: Weak Dominance

We must be careful when we use iterated elimination of *strictly* dominated strategies. To illustrate a potential pitfall, consider this game:

	Left	Right
Up	0, 1	-4, 2
Middle	0, 3	3, 3
Down	-2, 2	3, -1

Let's first focus on player 1's choice between up and middle:

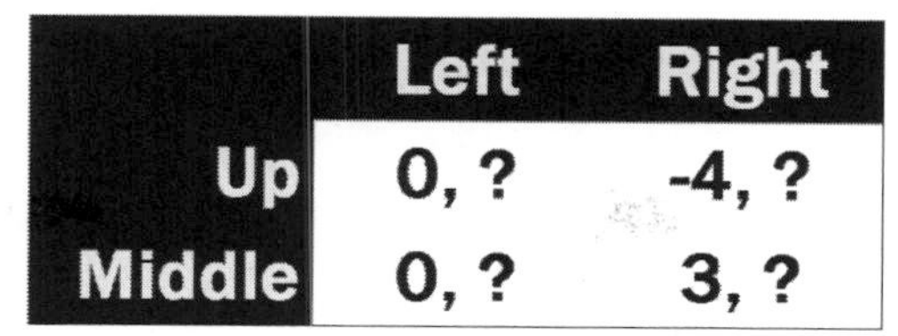

	Left	Right
Up	0, ?	-4, ?
Middle	0, ?	3, ?

You might be tempted to say that middle strictly dominates up for player 1. However, that is not the case. If player 2 plays right, then middle defeats up, as 3 is greater than -4. But if player 2 plays left, player 1 is indifferent between up and middle; regardless of which he chooses, he winds up with a payoff of 0. Strict dominance requires middle to *always* be better than up; equally good does not cut it.

Instead, we say that middle *weakly dominates* up. In general, a strategy x weakly dominates a strategy y for a player if x provides *at least* as great of a payoff for that player regardless of what the other players do and there is at least one set of opposing strategies for which x pays greater than y.

Eliminating weakly dominated strategies and analyzing the remaining game is called iterated elimination of weakly dominated strategies (IEWDS). Depending on the game, IEWDS sometimes produces sensible answers and sometimes does not. Unfortunately, we do not know which type of game we are looking at simply by eliminating weakly dominated strategies.

To fully see the problem, let's assume we could use IEWDS the same way we use IESDS. Since middle weakly dominated up, let's eliminate up and see what happens:

	Left	Right
Middle	0, 3	3, 3
Down	-2, 2	3, -1

We now see that left weakly dominates right for player 2:

	Left	Right
Middle	?, 3	?, 3
Down	?, 2	?, -1

That is, 3 is equal to 3 and 2 is greater than -1. If we eliminate the weakly dominated right strategy, this game remains:

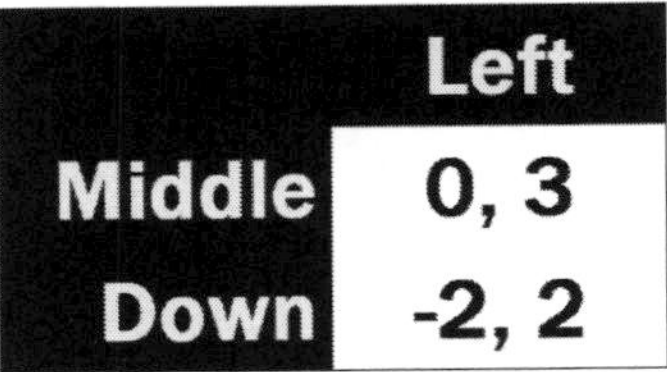

	Left
Middle	0, 3
Down	-2, 2

Since player 1 earns 0 for middle and -2 for down, he optimally chooses middle. Thus, we conclude that the outcome of this game is <middle, left>.

But is it? Let's start over:

	Left	Right
Up	0, 1	-4, 2
Middle	0, 3	3, 3
Down	-2, 2	3, -1

We began last time by observing that middle weakly dominates up. However, middle also weakly dominates down:

	Left	Right
Middle	0, ?	3, ?
Down	-2, ?	3, ?

That is, 0 is greater than -2 and 3 is equal to 3. Eliminating down yields the following:

	Left	Right
Up	0, 1	-4, 2
Middle	0, 3	3, 3

Now we see that right weakly dominates left for player 2:

	Left	Right
Up	?, 1	?, 2
Middle	?, 3	?, 3

That is, 2 is greater than 1 and 3 is equal to 3. So eliminating left leaves us with this:

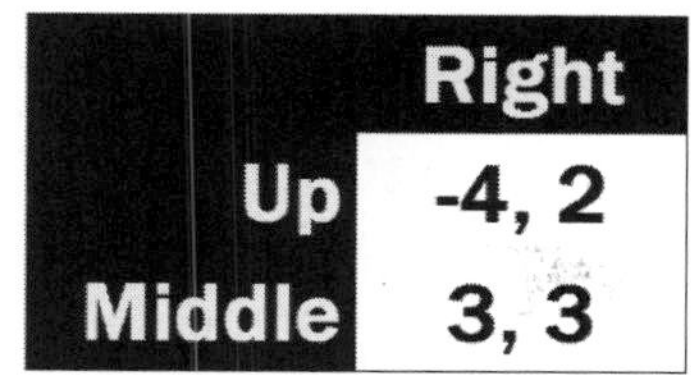

	Right
Up	-4, 2
Middle	3, 3

At this point, player 1 picks the strategy that earns him the highest payoff. Since 3 is greater than -4, he selects middle, leaving us with a solution of <middle, right>. Yet earlier we said the solution was <middle, left>. Depending on the order of elimination, IEWDS produces two separate answers.

The problem is that iterated elimination of weakly dominated strategies gives us no guidance about which is correct or if both are. To resolve the issue, we need to introduce a broader solution concept. We will do that in the next lesson.

Takeaway Points

1) Iterated elimination of strictly dominated strategies (IESDS) simplifies games by removing strictly dominated strategies—strategies that players would never play. This allows players to make inferences based on what others will *not* play.
2) Order does not matter for IESDS.
3) Strategy x *weakly dominates* strategy y for a player if x provides *at least* as great of a payoff for that player regardless of what the other players do and there is at least one set of opposing strategies for which x pays greater than y.
4) Iterated elimination of weakly dominated strategies (IEWDS) sometimes produces multiple answers.
5) The order of elimination matters for IEWDS.

Lesson 1.3: The Stag Hunt, Pure Strategy Nash Equilibrium, and Best Responses

Two hunters enter a range filled with hares and a single stag. Hares are unintelligent and easy to capture. The stag, on the other hand, is cunning—the hunters can only catch it by working together.

Without any communication, the hunters independently choose whether to hunt hares or the stag. If they both hunt hares, they each capture half of the hares in the range. If one hunts the stag and the other hunts hares, the stag hunter goes home empty-handed while the hare hunter captures all of the hares. Finally, if both hunt the stag, then each of their shares of the stag is greater than the value of all of the hares.

The following matrix depicts the strategic situation:

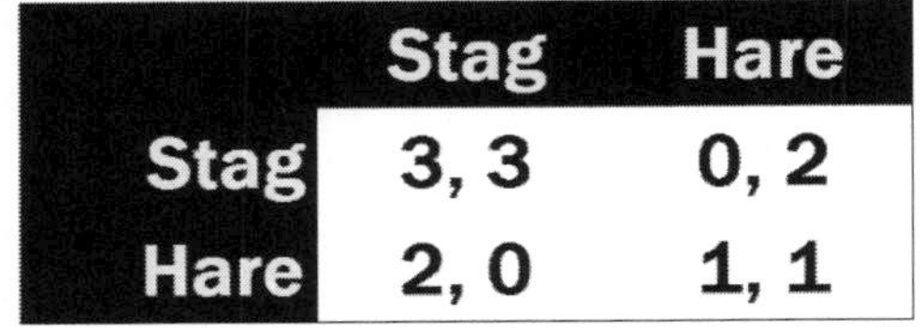

	Stag	Hare
Stag	3, 3	0, 2
Hare	2, 0	1, 1

Each player most prefers the <stag, stag> outcome. From that, we might assume that the <stag, stag> is the only sensible outcome of the game. However, the players could rationally wind up at a different outcome.

So far, we only know how to solve games using iterated elimination of strictly dominated strategies. Let's search for dominant strategies here.

First, suppose player 1 knew that player 2 will hunt the stag:

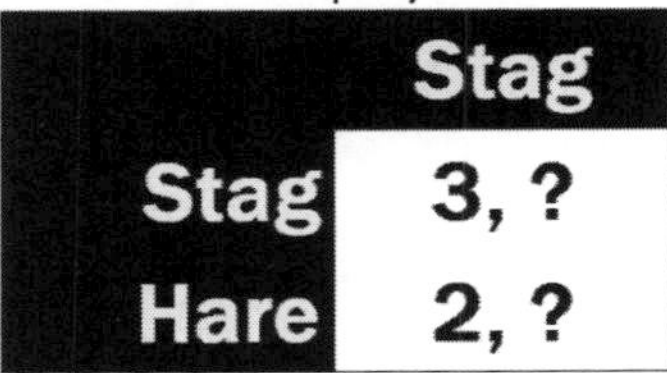

	Stag
Stag	3, ?
Hare	2, ?

In this case, hunting the stag is optimal for player 1 as well: doing so nets him 3, whereas chasing hares gives him only 2.

Now suppose player 1 knew player 2 will hunt hares:

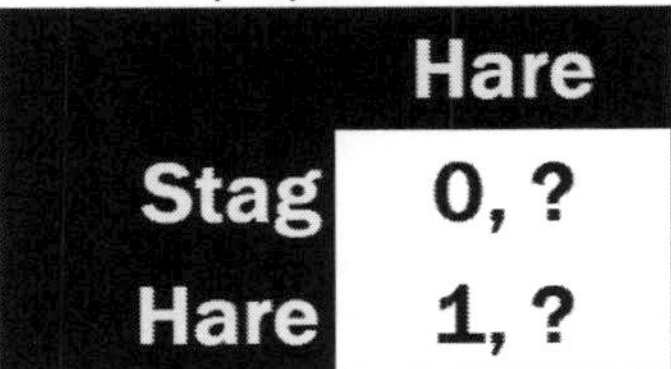

	Hare
Stag	0, ?
Hare	1, ?

Hunting the stag is no longer optimal for player 1; hare has stag beat by a 1 to 0 margin. Thus, player 1 has no strictly (or weakly) dominated strategy. In fact, player 1's optimal strategy is completely dependent on what player 2 selects. If she hunts the stag, so should he; but if she hunts for hares, he ought to as well.

Given the symmetry of the game, the same is true for player 2: she should also play whichever strategy player 1 selects.

To verify this, suppose player 1 hunted the stag:

	Stag	Hare
Stag	?, 3	?, 2

Player 2 should choose stag, since 3 beats 2.

But suppose player 1 hunted for hares:

	Stag	Hare
Hare	?, 0	?, 1

Then she ought to opt for hares, as 1 beats 0.

How do we solve games lacking dominated strategies? We look for *Nash equilibria*. A Nash equilibrium is a set of strategies, one for each player, such that no player has incentive to change his or her strategy given what the other players are doing.

Some examples will clarify that definition. First, consider the set of strategies <stag, stag>. Does either player have incentive to change his or her strategy?

Let's look at this from the perspective of player 1. First, we have to hold player 2's strategy constant; that is, we assume that player 2 will stick to her strategy of stag. Should player 1 switch his strategy?

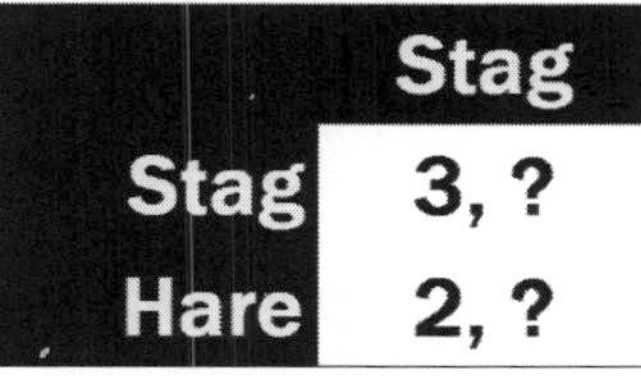

	Stag
Stag	3, ?
Hare	2, ?

We just saw this image, so we know he should not: 3 still beats 2.

What about player 2? Similarly, we must hold player 1's strategy constant and ask whether player 2 would want to deviate from her strategy:

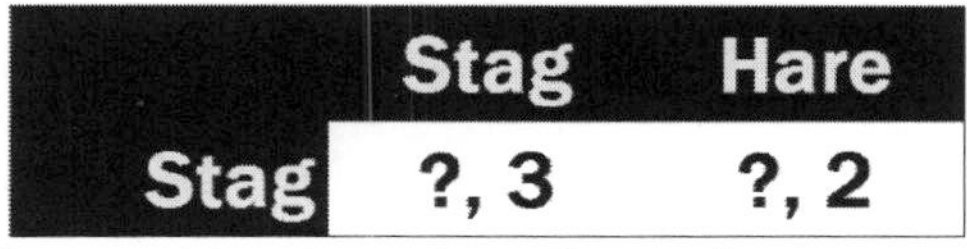

	Stag	Hare
Stag	?, 3	?, 2

Once more, she would not want to: 3 remains greater than 2. So <stag, stag> is a Nash equilibrium. Specifically, we call it a pure strategy Nash equilibrium (PSNE) because both players are playing deterministic strategies. That is, in this equilibrium, player 1 *always* plays stag and player 2 *always* plays stag. We will only focus on PSNE in this section. (In contrast, Lesson 1.5 covers mixed strategy Nash equilibrium, or MSNE, in which players randomize between their strategies. For example, a mixed strategy might be to flip a coin and hunt hares on heads and the stag on tails.)

Are there any other PSNE? Let's start by seeing if player 1 would want to switch his strategy in the <stag, hare> outcome:

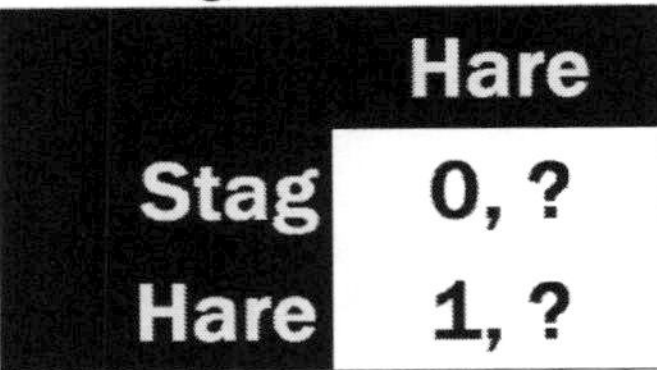

	Hare
Stag	0, ?
Hare	1, ?

He should alter his strategy. If he keeps hunting a stag, he will end up with 0. But if he switches his strategy to hare, he can profitably deviate to 1.

If even a single player would want to deviate, a set of strategies is not a Nash equilibrium. So without even checking player 2's move, we can throw out <stag, hare> as an equilibrium candidate. But let's check anyway for the sake of practice:

	Stag	Hare
Stag	?, 3	?, 2

This is the third time we have seen this image, so by now we should know player 2 has a profitable deviation: she should switch to stag.

Now let's check whether <hare, stag> is a Nash equilibrium. Given the game's symmetry and how <stag, hare> is not a Nash equilibrium, it should be obvious that <hare, stag> is not either. To wit, consider player 1's choice:

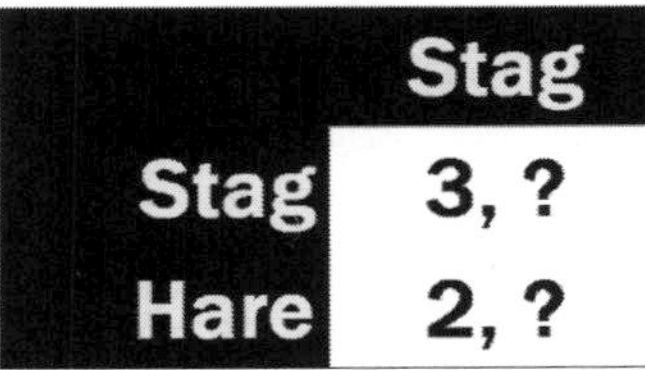

	Stag
Stag	3, ?
Hare	2, ?

Currently, he earns 2; if he switches to stag, he receives 3. Since that is a profitable deviation, <hare, stag> is not a Nash equilibrium.

Once more, we could have also verified that <hare, stag> is not a Nash equilibrium by looking at player 2's choice:

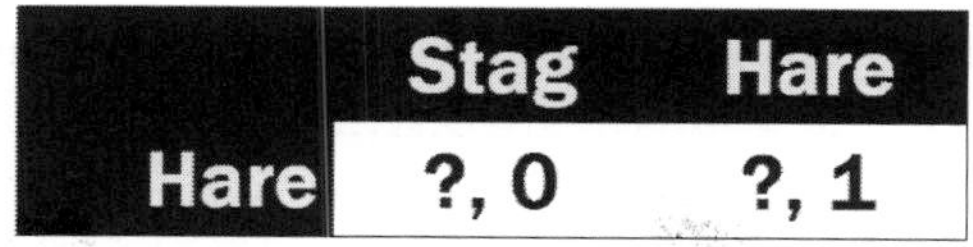

	Stag	Hare
Hare	?, 0	?, 1

Optimally, player 2 should switch from hunting the stag and earning 0 to chasing hares and earning 1.

Finally, let's check whether <hare, hare> is a Nash equilibrium. We will begin with player 1's choice:

	Hare
Stag	0, ?
Hare	1, ?

Hare remains optimal for player 1; switching to stag decreases his payoff from 1 to 0. So the only way for <hare, hare> to not be a Nash equilibrium is if player 2 would want to switch. Let's check if that is the case:

	Stag	Hare
Hare	?, 0	?, 1

She should not switch—deviating also decreases her payoff from 1 to 0. Since neither player has incentive to change his or her strategy, <hare, hare> is a Nash equilibrium. Therefore, the stag hunt has two pure strategy Nash equilibria: <stag, stag> and <hare, hare>.

Unlike the prisoner's dilemma, the stag hunt illustrates game theory's power to analyze interdependent decision making. In the prisoner's dilemma, each player could effectively ignore what the other one planned on doing since confess generated a strictly greater payoff regardless of the other prisoner's choice. That is not the case with the stag hunt. Here, each player wants to do what the other is doing. That is, each player's individually optimal strategy is a function of the other player's choice.

The stag hunt also highlights how Nash equilibria do not have to be efficient. Indeed, both players are better off with the <stag, stag> equilibrium than the <hare, hare> equilibrium—and, unlike the prisoner's dilemma, both outcomes are sustainable, because neither player has incentive to change his or her strategy given what the other is doing. As such, although the players might want to coordinate on the <stag, stag> outcome, they might get stuck in the inefficient <hare, hare> outcome anyway.

To see how this is plausible, suppose both players saw a sign that read "today is hare hunting day" on their way to the hunting range. What should the hunters do? Unfortunately, the structure of the game provides no clear answer. It is in each hunter's best interest to ignore the sign and hunt a stag, as long as both do that. But if I think that you are going to follow the sign and hunt hares, then I should follow the sign as well, even if I really want to hunt a stag. And if you think I am going to think that you are going to follow the sign, even if you have no plans to do so, you should still hunt hares because you anticipate that I will as well. Simple coordination—a cell phone call from one honest hunter to the other—could solve the issue. But absent that, both outcomes are plausible, which is one of the motivations behind Nash equilibria.

As a final note, Nash equilibrium only looks at *individual* deviations. That is, we need to check whether individuals cannot individually deviate to better payoffs. If both players are playing hare, then there is a collectively profitable deviation to both playing stag. But individually, both are better off staying with their hare strategy. In this sense, Nash equilibrium has a "no regrets" property. If players play according to a Nash equilibrium, then they do not regret their choices once they have realized their payoffs.

1.3.1: New Preferences for the Prisoner's Dilemma

Lesson 1.1 claimed that if players did not solely want to minimize jail time, their optimal behaviors might change. Indeed, game theory was *not* making a normative claim that jail time is the only thing that players should care about; instead, we were seeking the optimal behaviors of the prisoners given that they had such preferences.

With that in mind, let's reframe the game. Suppose both prisoners were good friends and would rather keep quiet if they knew that the other would as well. In this case, each player's most preferred outcome is <quiet, quiet>, then <confess, quiet>, then <confess, confess>, then <quiet, confess>. Payoff values of 3, 2, 1, and 0 preserve the rankings of these outcomes. Consequently, we could draw the payoff matrix like this:

	Quiet	Confess
Quiet	3, 3	0, 2
Confess	2, 0	1, 1

Notice that these are the exact same payoffs that we saw in the stag hunt. As such, there are two PSNE: <keep quiet, keep quiet> and <confess, confess>. Now the interrogator's plan may fail, as the players have a strong enough relationship that their preferences allow for mutual cooperation. That is, <keep quiet, keep quiet> is a sustainable outcome in this version of the prisoner's dilemma, as neither player has incentive to deviate from that set of strategies.

This reworking of the prisoner's dilemma provides two takeaways. First, it shows how outcomes are a function of preferences and not just the strategic environment. The interrogator made the same offer in both versions of the game. In the original version, he induced both to confess. But when the players had friendlier preferences, his plot could very well fail, provided the prisoners can coordinate on the <quiet, quiet> equilibrium.

Second, it once again highlights game theory's ability to draw parallels between seemingly dissimilar situations. In this section alone, we have looked at two completely different scenarios. One dealt with hunting; the other dealt with two friends in a legal predicament. Yet, once we strip down all of the irrelevant features of the strategic interaction, we see that a single game underlies both cases. In this way, game theory allows us to connect seemingly disparate situations under a common framework.

1.3.2: Safety in Numbers and Best Responses

Two generals each have three units and are preparing for an upcoming battle. Each can choose to send any number of units to the fight or none at all. The side with more troops wins the battle, and the fight will draw if there are equal forces. Victory is worth 1 point; defeat is worth -1. If the sides draw or at least one side refuses to fight, both sides earn 0.

This is a "safety in numbers" game, and the following matrix represents the generals' situation:

	Zero	One	Two	Three
Zero	0, 0	0, 0	0, 0	0, 0
One	0, 0	0, 0	-1, 1	-1, 1
Two	0, 0	1, -1	0, 0	-1, 1
Three	0, 0	1, -1	1, -1	0, 0

When we found all the pure strategy Nash equilibria in the stag hunt, we went through each outcome one at a time and checked whether a player had a profitable deviation. While that was a reasonable task when there were only four outcomes, safety in numbers has sixteen. We could go through each of the sixteen outcomes, but that would be time consuming.

Instead, we will use a different method that involves marking best responses.

A *best response* is simply the optimal strategy for a particular player given what everyone else is doing. For example, suppose general 2 was going to send 0 units to the battle. What is general 1's best response?

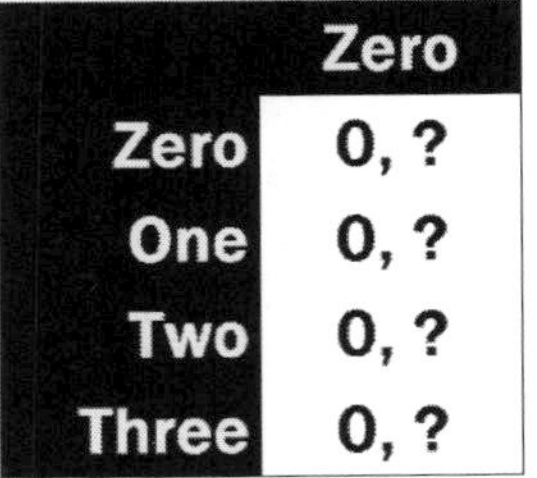

	Zero
Zero	0, ?
One	0, ?
Two	0, ?
Three	0, ?

As the image demonstrates, general 1's strategy becomes irrelevant—the battle never takes place, so every choice generates a payoff of 0 for him. As such, *all* of these strategies are best responses for general 1. For bookkeeping purposes, we mark a player's best responses with an asterisk over his or her payoffs, like this:

	Zero
Zero	0*, ?
One	0*, ?
Two	0*, ?
Three	0*, ?

We will see why these asterisks come in handy in a moment.

Now suppose general 2 will send one unit to the battle. Let's mark the first general's best responses:

	One
Zero	0, ?
One	0, ?
Two	1*, ?
Three	1*, ?

Two best responses exist: sending two and sending three units. Sending zero units results in no battle taking place and a payoff of 0. Sending one unit results in a draw and a payoff of 0. Sending two or sending three units gives general 1 the victory and a payoff of 1. Since 1 is the greatest payoff, we mark two and three as best responses.

Next, suppose general 2 sent two units to the battle:

	Two
Zero	0, ?
One	-1, ?
Two	0, ?
Three	1*, ?

Here, sending one loses the battle, sending two leads to a draw, and sending three earns the win. Since the payoff of 1 is the greatest of all the outcomes, general 1's best response is to send three units. As such, only three gets an asterisk.

Lastly, suppose general 2 sent three units:

	Three
Zero	0*, ?
One	-1, ?
Two	-1, ?
Three	0*, ?

Now there are two best responses. If general 1 sends one or two units, he loses the battle and earns -1. If he sends no one, the battle does not take place, and he gets 0. And if he sends everyone, the battle draws, and he also receives a payoff of 0. Both of the 0 outcomes are the best, so both earn asterisks.

Having marked all of general 1's best responses, the game looks like this:

	Zero	One	Two	Three
Zero	0*, 0	0, 0	0, 0	0*, 0
One	0*, 0	0, 0	-1, 1	-1, 1
Two	0*, 0	1*, -1	0, 0	-1, 1
Three	0*, 0	1*, -1	1*, -1	0*, 0

However, we are only half way done—we still need to mark general 2's best responses. As before, let's start by assuming general 1 will send zero units:

	Zero	One	Two	Three
Zero	?, 0*	?, 0*	?, 0*	?, 0*

We see that every strategy is a best response for general 2 if general 1 sends zero units.

Now check what happens if general 2 sends one unit:

	Zero	One	Two	Three
One	?, 0	?, 0	?, 1*	?, 1*

Sending two or three units nets 1. Therefore, two and three are best responses.

Next, let's consider general 2's best responses to general 1 sending two units:

	Zero	One	Two	Three
Two	?, 0	?, -1	?, 0	?, 1*

Sending three units nets 1 point, whereas the other alternatives are worth 0 or -1. So sending three units is the only best response here.

Finally, suppose general 1 sent three units:

	Zero	One	Two	Three
Three	?, 0*	?, -1	?, -1	?, 0*

Now one and two are losing strategies, while zero and three break even at 0. So zero and three are best responses.

Combining all of that information together, the full game looks like this:

	Zero	One	Two	Three
Zero	0*, 0*	0, 0*	0, 0*	0*, 0*
One	0*, 0	0, 0	-1, 1*	-1, 1*
Two	0*, 0	1*, -1	0, 0	-1, 1*
Three	0*, 0*	1*, -1	1*, -1	0*, 0*

Was all of that work worth the effort? Most definitely. To find all of the game's pure strategy Nash equilibria, we only need to check which outcomes have asterisks next to both players' payoffs. We see there are four such outcomes: <0, 0>, <0, 3>, <3, 0>, and <3, 3>.

Why are these outcomes Nash equilibria? Recall that a Nash equilibrium is a set of strategies, one for each player, such that no player has incentive to change his or her strategy given what the other players are doing. If an outcome is a best response for a player, he or she cannot change strategies and earn a greater payoff. But if the outcome is a best response for all players, no player has incentive to change his or her strategy. That matches our definition of Nash equilibrium.

As such, an alternative definition of Nash equilibrium is a mutual best response. This equivalent definition can come in handy depending on the game in question.

1.3.3: The Stoplight Game

We have seen the definition of Nash equilibrium many times now. But what exactly does it mean? One interpretation is that a Nash equilibrium is a law that everyone would want to follow even in the absence of an effective police force.

For example, consider the role of stoplights in a society. Imagine two cars are approaching an intersection at 40 miles per hour from perpendicular directions. If both continue full speed, they will crash spectacularly. But if both stop, they waste time deciding who should go through the intersection first. Both drivers benefit if one continues without stopping while the other momentarily brakes to allow the other to pass.

We can illustrate the drivers' choices and preferences using the following matrix:

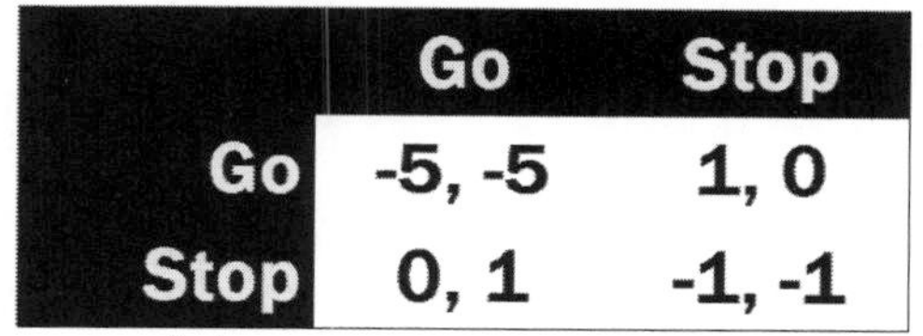

	Go	Stop
Go	-5, -5	1, 0
Stop	0, 1	-1, -1

Let's mark the game's best responses. First, suppose player 2 chose go. How should player 1 respond?

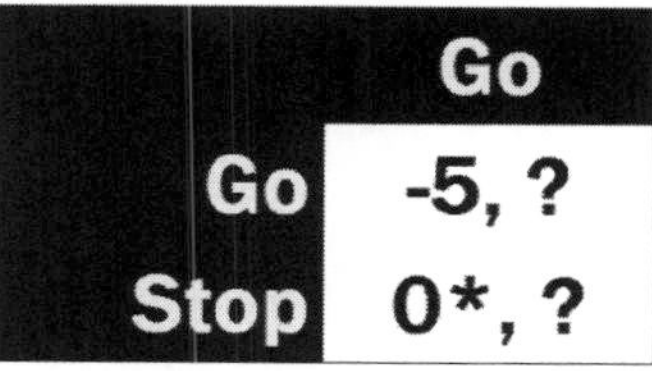

	Go
Go	-5, ?
Stop	0*, ?

Stopping is optimal. If player 1 goes as well, he causes a devastating accident. Although stopping slows player 1 down, at least he will make it to his destination alive.

Now suppose player 2 picked stop instead:

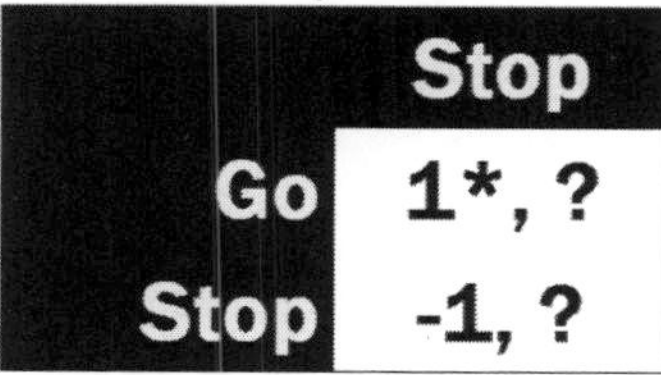

	Stop
Go	1*, ?
Stop	-1, ?

Stopping is unnecessary for player 1, as it results in an awkward moment where the drivers try to decide who ought to drive off first. Meanwhile, if player 1 goes, he arrives at his destination as quickly as possible. As such, the 1 for go receives the asterisk.

The story is the same for player 2's best responses. If player 1 goes, player 2 ought to stop:

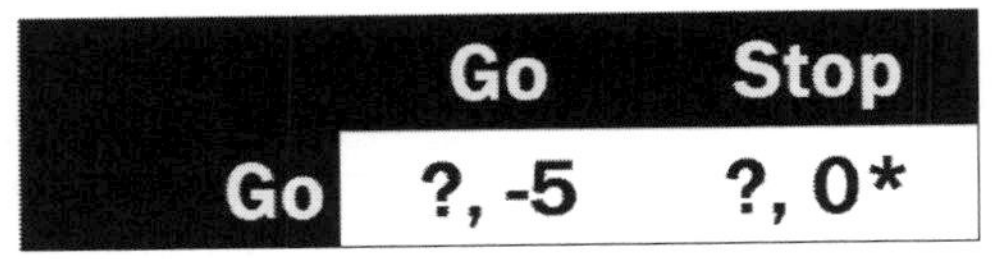

	Go	Stop
Go	?, -5	?, 0*

After all, 0 is greater than -5.

But if player 1 stops, player 2 should go:

	Go	Stop
Stop	?, 1*	?, -1

This time, 1 is greater than -1.

If we fold these best responses together, the original game looks like this:

	Go	Stop
Go	-5, -5	1*, 0*
Stop	0*, 1*	-1, -1

So the game has two pure strategy Nash equilibria: <go, stop> and <stop, go>. While these Nash equilibria leave both players better off than the <go, go> or <stop, stop> outcomes, player 1 prefers the <go, stop> equilibrium while player 2 prefers the <stop, go> equilibrium. Consequently, coordination is not straightforward; player 1 wants his version of coordination, while player 2 wants hers.

How can the players resolve their dilemma? Stoplights provide a solution. The stoplight tells one driver to go with a green light, while it orders the other to stop with a red light. The players have no incentive to deviate from the stoplight's suggestion. If the driver at the red light goes, he causes an accident. If the driver at the green light stops, he unnecessarily wastes some time. Thus, the stoplight instructs the drivers to play a Nash equilibrium.

Note that these strategies are self-reinforcing. One driver wants to stop because he knows the other driver will go. Likewise, the other driver wants to go because he knows the first driver will stop. The players do not need a third party—say, police officers—to enforce the equilibrium. Instead, each player naturally wants to execute his intended strategy because the other player's strategy makes it optimal for him to do so. Again, this is due to the no regrets property of Nash equilibrium. Drivers follow the stoplight because they know they will not regret doing so afterward.

Conclusion

Since the best responses method is an efficient way of to find pure strategy Nash equilibria, we will be relying on it in upcoming lessons. Consequently, if you do not yet have the hang of it, there will be plenty more practice ahead.

Takeaway Points

1) A *Nash equilibrium* is a set of strategies, one for each player, such that no player has incentive to change his or her strategy given what the other players are doing.
2) A player's *best response* is the strategy or strategies that produce the greatest payoff given all other players' strategies.
3) We can find pure strategy Nash equilibrium by marking each player's best responses in a game matrix. Outcomes that are best responses for all players are Nash equilibria.
4) Nash equilibria can be thought of as laws that no one would want to break even in the absence of an effective police force.
5) Nash equilibria have a "no regrets" property—after the game has been played, players do not regret their choices if all were playing a Nash equilibrium.

Lesson 1.4: Dominance and Nash Equilibrium

This lesson investigates the relationship between Nash equilibrium and dominance. We will see two things. First, if iterated elimination of strictly dominated strategies reduces the game to a single outcome, that outcome is a Nash equilibrium and it is the *only* Nash equilibrium of that game. Meanwhile, iterated elimination of weakly dominated strategies is not as kind: although any solution found through IEWDS is a Nash equilibrium, the IEWDS process sometimes eliminates other Nash equilibria.

1.4.1: Nash Equilibrium and Iterated Elimination of Strictly Dominated Strategies

How does Nash equilibrium relate to iterated elimination of strictly dominated strategies? As it turns out, solutions found through the strict dominance solution concept are Nash equilibria. Consequently, not only is <confess, confess> the dominance solution to the prisoner's dilemma from Lesson 1.1, but it is also the Nash equilibrium of that game.

Let's consider a tougher example. Here is the game from Lesson 1.2 we used to introduce iterated elimination of strictly dominated strategies:

	Left	Center	Right
Up	13, 3	1, 4	7, 3
Middle	4, 1	3, 3	6, 2
Down	-1, 9	2, 8	8, -1

Through IESDS, we know the solution is <middle, center>. Let's verify that it is also the unique pure strategy Nash equilibrium of the game.

First, let's find player 1's best response to left:

	Left
Up	13*, ?
Middle	4, ?
Down	-1, ?

Up earns the asterisk, as 13 is the largest payoff.

Moving on, suppose player 2 chose center:

	Center
Up	1, ?
Middle	3*, ?
Down	2, ?

Middle gets the asterisk this time, as 3 is now the largest payoff for player 1.

Lastly for player 1's best responses, we consider player 2 moving right:

	Right
Up	7, ?
Middle	6, ?
Down	8*, ?

Down earns the asterisk with a payoff of 8.

Switching gears, let's find the best responses for player 2 if player 1 moves up:

	Left	Center	Right
Up	?, 3	?, 4*	?, 3

With 4 being the largest, player 2's best response is to move center.

Now suppose player 1 moved middle:

	Left	Center	Right
Middle	?, 1	?, 3*	?, 2

Center remains the best response, with 3 being greater than 1 or 2.

Finally, consider how player 2 should reply to down:

	Left	Center	Right
Down	?, 9*	?, 8	?, -1

Player 2's best response has shifted to left.

If we reconstruct the original game with all of the best responses marked, we get this:

	Left	Center	Right
Up	13*, 3	1, 4*	7, 3
Middle	4, 1	3*, 3*	6, 2
Down	-1, 9*	2, 8	8*, -1

The only mutual best response is <middle, center>. As claimed, the solution IESDS is the only Nash equilibrium of the game.

Why does this work? Recall that Nash equilibrium strategies have the "no regrets" property. Yet players immediately regret having played strictly dominated strategies, since they could have performed better by selecting the dominating strategy. As a result, if IESDS produces a single outcome, the strategies associated with that outcome are the only strategies that players will not regret having selected. In turn, those strategies form the only Nash equilibrium.

1.4.2: When IESDS Leaves Multiple Strategies

After eliminating all strictly dominated strategies, sometimes multiple strategies remain available for each player. How do we handle this? We simply fall back to the other techniques to find Nash equilibria.

To illustrate, let's add some spice to the stag hunt. This time, player 1 can hunt a stag, hares, or player 2. To keep player 2 from being defenseless, she can hunt a stag, hunt hares, or hide from player 1. If player 1 goes after his hunting partner and player 2 hides, then player 2 survives. Otherwise, she earns an extremely negative payoff. (She will be shot, after all.) Fortunately for player 2, player 1 will feel guilty if he ever shoots player 2, so he also earns a negative payoff if she is ever hurt.

Adding these new strategies and payoffs leaves us with this game:

	Stag	Hare	Hide
Stag	3, 3	0, 2	0, 0
Hare	2, 0	1, 1	2, 0
Human	-5, -10	-5, -10	0, 0

To solve this game, we should first note that hare strictly dominates human for player 1:

	Stag	Hare	Hide
Hare	2, ?	1, ?	2, ?
Human	-5, ?	-5, ?	0, ?

That is, 2 beats -5, 1 beats -5, and 2 beats 0. Knowing that, we can remove human as a strategy and consider the following:

	Stag	Hare	Hide
Stag	3, 3	0, 2	0, 0
Hare	2, 0	1, 1	2, 0

From here, note that hare strictly dominates hide for player 2:

	Hare	Hide
Stag	?, 2	?, 0
Hare	?, 1	?, 0

This is because 2 is greater than 0 and 1 is also greater than 0. Essentially, player 2 has no reason to hide if she knows player 1 would never hunt her.

If we remove hide, the game reduces to the original stag hunt:

	Stag	Hare
Stag	3, 3	0, 2
Hare	2, 0	1, 1

From Lesson 1.3, we know no more dominated strategies remain. To solve the game, we simply look for Nash equilibria as before. The two mutual best responses in pure strategies are <stag, stag> and <hare, hare>, so those are the two pure strategy Nash equilibria of this game. More pertinent to this topic, however, they are also the only two PSNE of the original game with human hunting and hiding.

1.4.3: Nash Equilibrium and Iterated Elimination of Weakly Dominated Strategies

Previously, we looked at this game:

	Left	Right
Up	0, 1	-4, 2
Middle	0, 3	3, 3
Down	-2, 2	3, -1

Middle weakly dominates both up and down for player 1. Depending on whether we chose to eliminate up or down first as a part of a sequence of iterated elimination of weakly dominated strategies, either <middle, left> or <middle, right> was the only outcome that remained. So which of these is a pure strategy Nash equilibrium? Actually, *both* are.

As always, let's mark the best responses for both players. Begin by checking what happens if player 1 plays up:

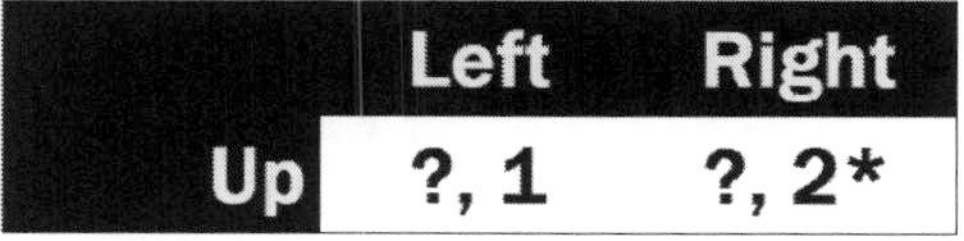

	Left	Right
Up	?, 1	?, 2*

Since 2 beats 1, right is player 2's best response.

Now check middle:

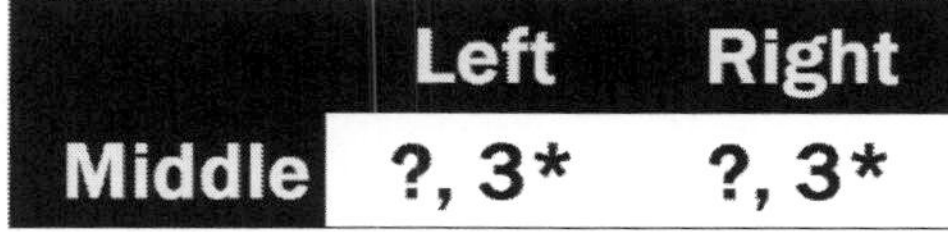

	Left	Right
Middle	?, 3*	?, 3*

If player 1 plays middle, player 2 is indifferent between left and right; both are worth 3 to her. As such, both are best responses.

Lastly, we go to down:

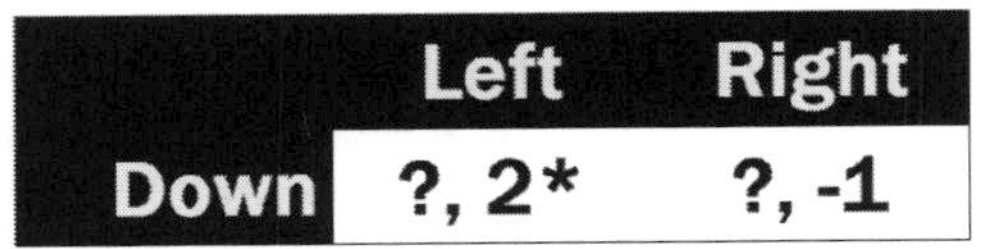

	Left	Right
Down	?, 2*	?, -1

Left is the best response here, as 2 beats -1.

From here, we move to player 1's best responses, beginning with player 2 moving left:

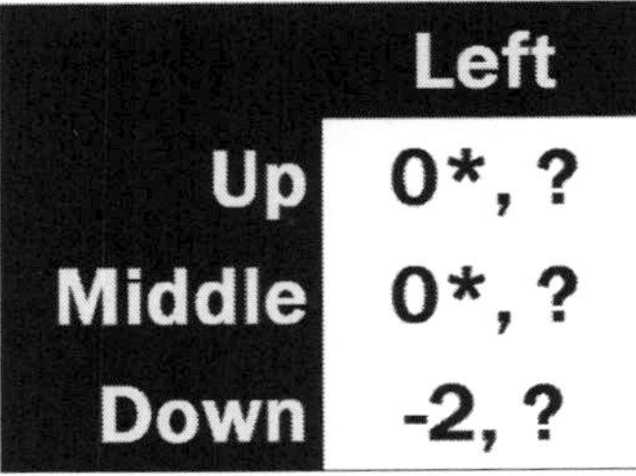

	Left
Up	0*, ?
Middle	0*, ?
Down	-2, ?

Up and middle are both best response to left; they are both worth 0, while down yields -2.

The final situation to consider is player 2 moving right:

	Right
Up	-4, ?
Middle	3*, ?
Down	3*, ?

Again, player 1 has two best responses: middle and down.

Thus, the full game looks like this:

	Left	Right
Up	0*, 1	-4, 2*
Middle	0*, 3*	3*, 3*
Down	-2, 2*	3*, -1

We see that both <middle, left> and <middle, right> are mutual best responses. Therefore, both are pure strategy Nash equilibria.

This game only offers a hint of why it is not a good idea to exclusively rely on weakly dominated strategies. After all, as long as you had considered every alternative sequence IEWDS for this game, you would have found all the Nash equilibria. But weak dominance is much more finicky than that. In fact, even if there is only one path of elimination, IEWDS may completely erase a Nash equilibrium.

Let's use this game as an example:

	Left	Right
Up	2, 3	4, 3
Down	3, 3	1, 1

The only dominance here is involves left and right for player 2:

	Left	Right
Up	?, 3	?, 3
Down	?, 3	?, 1

Specifically, left weakly dominates right. That is, if player 1 plays up, she is equally well off playing left or right, but she does strictly better playing left if he moves down.

If we remove right from the game, this game remains:

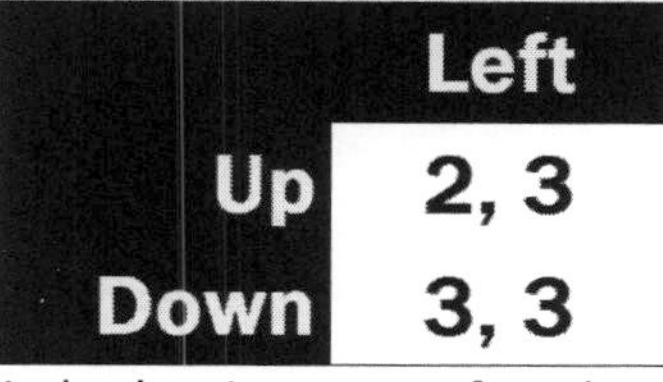

	Left
Up	2, 3
Down	3, 3

Since down now strictly dominates up for player 1, <down, left> is the solution that remains and it was also the *only* solution we can arrive at through iterated elimination of weakly dominated strategies.

Unfortunately, best responses tell a different story. Let's start with player 1's best response to player 2 playing left:

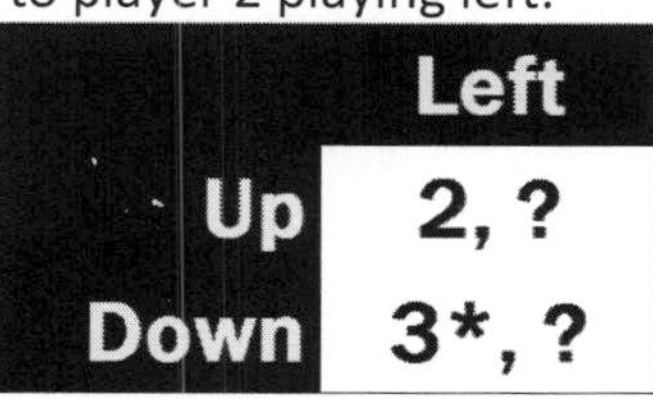

	Left
Up	2, ?
Down	3*, ?

As we just saw, it is down. Therefore, the 3 receives the asterisk.

Now check for player 1's best response to right:

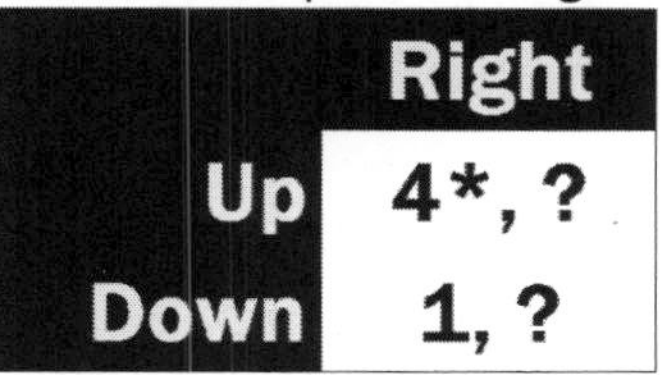

	Right
Up	4*, ?
Down	1, ?

Player 1's best response has flipped to up, as 4 beats 1.

Switching perspectives, let's mark player 2's best response to player 1 moving up:

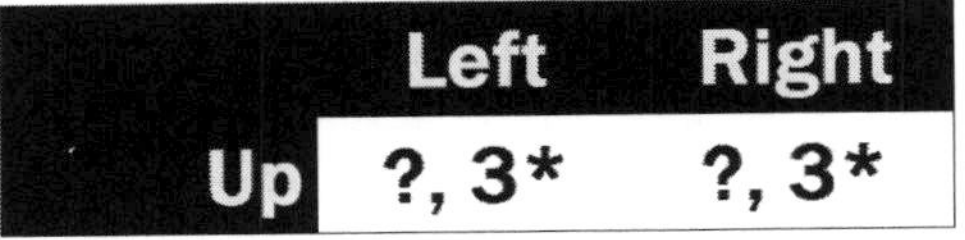

	Left	Right
Up	?, 3*	?, 3*

When player 1 goes up, player 2 is indifferent between left and right. As such, both are best responses.

Finally, let's check what happens when player 1 plays down:

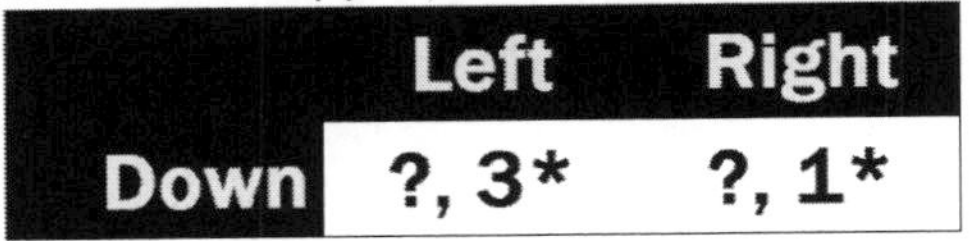

	Left	Right
Down	?, 3*	?, 1*

This time, left is the unique best response.

Let's bring all the best responses together:

	Left	Right
Up	2, 3*	4*, 3*
Down	3*, 3*	1, 1

From this, we see there are two pure strategy Nash equilibria: <up, right> and <down, left>. There is no way we could have used iterated elimination of weakly dominated strategies to arrive at <up, right>, which is why it is so dangerous to use IEWDS by itself.

On the other hand, sometimes IEWDS turns out just fine. To wit, consider this game:

	Left	Center	Right
Up	2, 2	4, 2	4, 3
Middle	2, 4	5, 5	7, 3
Down	3, 4	3, 7	6, 6

To start, note that middle weakly dominates up:

	Left	Center	Right
Up	2, ?	4, ?	4, ?
Middle	2, ?	5, ?	7, ?

Middle always produces at least as great of a payoff as up. When player 2 plays center, middle beats up by 5 to 4. When she plays right, middle again wins by 7 to 4. However, both up and middle are worth 2 when she selects left.

If we eliminated up as a strategy, the game reduces to this:

	Left	Center	Right
Middle	2, 4	5, 5	7, 3
Down	3, 4	3, 7	6, 6

And now center strictly dominates left for player 2:

	Left	Center
Middle	2, 4	5, 5
Down	3, 4	3, 7

If player 1 selects middle or down and player 2 chooses left, player 2 earns 4. In contrast, she earns 5 if player 1 plays middle and 7 if he plays down. So center strictly dominates left in the remaining game. Removing left yields the following:

	Center	Right
Middle	5, 5	7, 3
Down	3, 7	6, 6

Notice that center also strictly dominates right:

	Center	Right
Middle	?, 5	?, 3
Down	?, 7	?, 6

That is, 5 beats 3 and 7 beats 6. So we can remove right from the game:

	Center
Middle	5, 5
Down	3, 7

At this point, player 1 should pick middle and earn 5 rather than choose down and earn 3. Therefore, <middle, center> is the IEWDS solution and a pure strategy Nash equilibrium by extension.

Unlike previous examples, however, it is also the *unique* pure strategy Nash equilibrium of the game. We can verify this through best responses. Let's begin with player 1's choice if player 2 goes left:

	Left
Up	2, ?
Middle	2, ?
Down	3*, ?

Since 3 is the largest number there, down receives the asterisk.

Next, check for player 1's response to center:

	Center
Up	4, ?
Middle	5*, ?
Down	3, ?

Again, there is a single best response: middle with 5, beating up with 4 and down with 3.

For the final best response for player 1, we check what happens when player 2 plays right:

	Right
Up	4, ?
Middle	7*, ?
Down	6, ?

Middle wins once more with a 7.

Switching to the other side, let's find player 2's best response if player 1 goes up:

	Left	Center	Right
Up	?, 2	?, 2	?, 3*

Right wins, as it is worth 3 in response to up compared to the 2 for left or center.

Moving on, we now look at the best response to middle:

	Left	Center	Right
Middle	?, 4	?, 5*	?, 3

Here, the 5 for center beats out the 4 for left and the 3 for right. So center is player 2's best response to middle.

Finally, we go to the best response to down:

	Left	Center	Right
Down	?, 4	?, 7*	?, 6

For the second time, center is the best response.

Throwing all of that together, the whole game looks like this:

	Left	Center	Right
Up	2, 2	4, 2	4, 3*
Middle	2, 4	5*, 5*	7*, 3
Down	3*, 4	3, 7*	6, 6

We see there is a single mutual best response: <middle, center>. Fortuitously, this was also the answer iterated elimination of weakly dominated strategies gave us. For once, IEWDS actually functioned nicely.

You might wonder how you can tell whether IEWDS will work before you start solving the problem. Unfortunately, you cannot. It either will eliminate Nash equilibria or it will not, and you will not know which is the case until you check for best responses.

1.4.4: Simultaneous Strict and Weak Dominance

Occasionally, you will encounter a game that has both strictly and weakly dominated strategies in it. How should you go about solving the game? The most efficient answer is to begin by eliminating all of the strictly dominated strategies and only falling back on weak dominance when strict dominance cannot get you any further. The reason is simple: as soon as we eliminate a weakly dominated strategy—even just once—we may be eliminating some Nash equilibria.

To illustrate why you should eliminate strictly dominated strategies first, consider this game:

	Left	Right
Up	1, 1	2, -2
Middle	1, 0	1, -1
Down	0, 3	2, 2

Note that up weakly dominates middle for player 1:

	Left	Right
Up	1, ?	2, ?
Middle	1, ?	1, ?

That is, if player 2 plays left, player 1 is indifferent between up and middle; but if player 2 plays right, player 1 earns more from up than he does from middle. As such, up weakly dominates middle.

Removing middle from the matrix yields this:

	Left	Right
Up	1, 1	2, -2
Down	0, 3	2, 2

Now left strictly dominates right for player 2:

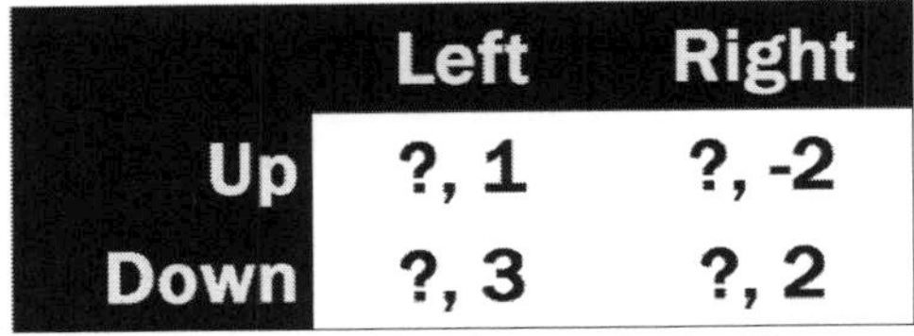

	Left	Right
Up	?, 1	?, -2
Down	?, 3	?, 2

This is because 1 beats -2 and 3 beats 2. So if we remove the strictly dominated strategy, we are left with the following:

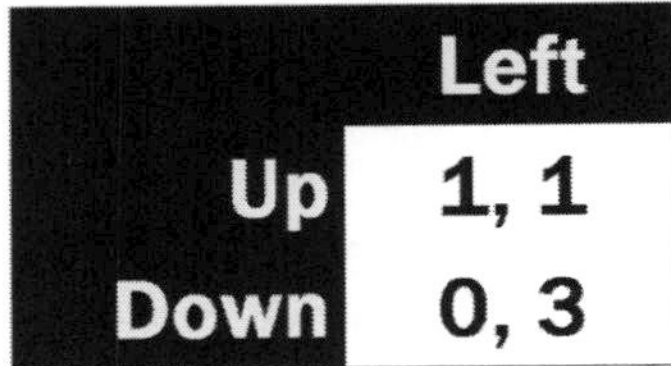

	Left
Up	1, 1
Down	0, 3

At this point, player 1 picks the strategy that gives him the greatest payoff, so he selects up. Thus, we have arrived at <up, left> being a solution to this game. Since a Nash equilibrium in a reduced game must also be a Nash equilibrium any of the previous games, <up, left> is a Nash equilibrium in the original game.

However, if we highlight the best responses of the original game, we find that the IEWDS process has eliminated another Nash equilibrium. To see this, consider player 1's best response to player 2 moving left:

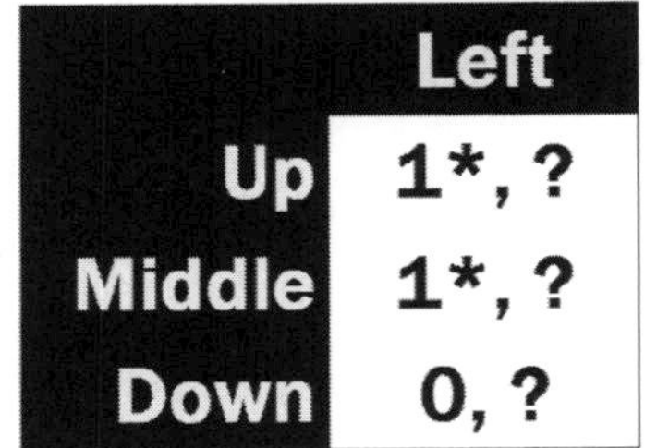

	Left
Up	1*, ?
Middle	1*, ?
Down	0, ?

There are two: up and middle. Both provide 1, whereas down provides 0. As such, up and middle are best responses, but down is not.

Now let's check for player 1's best response to right:

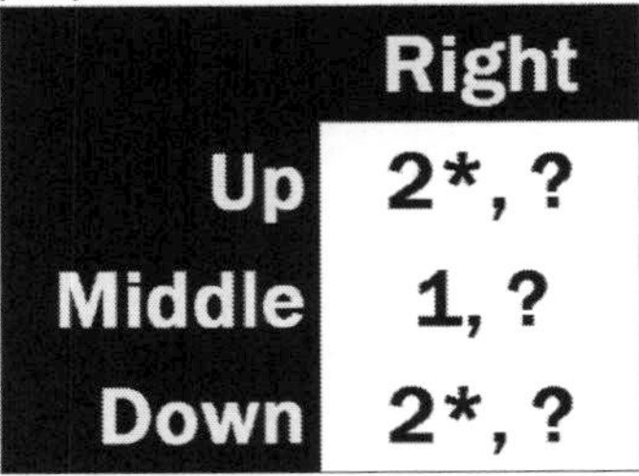

	Right
Up	2*, ?
Middle	1, ?
Down	2*, ?

Again, there are two: up and down. Middle loses out here by a score of 2 to 1.

Switching gears, we move to player 2's best response to up:

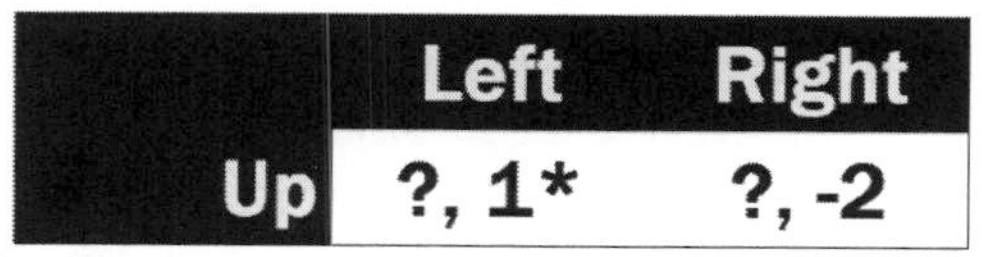

	Left	Right
Up	?, 1*	?, -2

Here, left is the best response, as 1 beats -2.

Moving on to middle:

	Left	Right
Middle	?, 0*	?, -1

Again, left wins, so it earns the asterisk.

Finally, we go to down:

	Left	Right
Down	?, 3*	?, 2

For a third time, left is the best response.

Putting all of those best responses together, this is what the original game looks like:

	Left	Right
Up	1*, 1*	2*, -2
Middle	1*, 0*	1, -1
Down	0, 3*	2*, 2

Weak dominance has failed us once more: IEWDS produced an answer of <up, left>, but both <up, left> and <middle, left> are mutual best responses and are therefore pure strategy Nash equilibria. Utilizing weak dominance would have yielded only a partially correct answer here.

However, you might have noticed in the process of marking best responses that left strictly dominates right—regardless of what player 1 does, player 2 earns more by picking left than she does by picking right. So instead of beginning the elimination sequence with up weakly dominating middle, we could have begun by removing right. Had we done so, we would have been looking at this game:

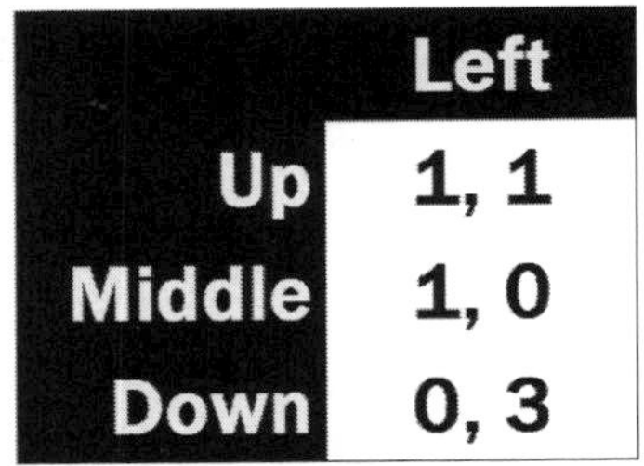

	Left
Up	1, 1
Middle	1, 0
Down	0, 3

At this point, up and middle both strictly dominate down, since 1 is greater than 0. Reducing the game one step further, we are left with only two outcomes:

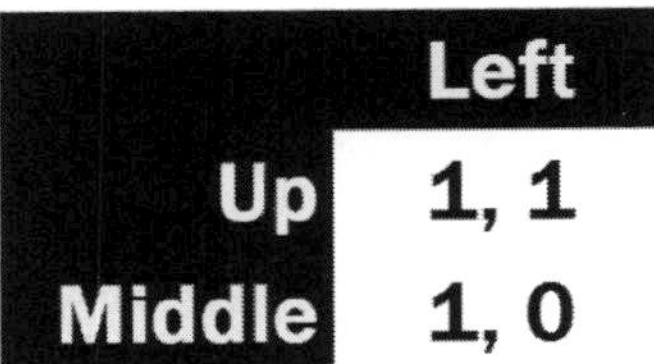

	Left
Up	1, 1
Middle	1, 0

Since both up and middle are best responses to left, both <up, left> and <middle, left> are the IESDS solutions. And as we already know, they are also the only pure strategy Nash equilibria of the original game. Consequently, IESDS has preserved all the Nash equilibria.

Attacking the strictly dominated strategies first side-stepped the weak dominance problem. After we removed right from the game, no instance of weak dominance remained, only more cases of strict dominance. Unfortunately, this will not always be the case: sometimes after removing all strictly dominated strategies weakly dominated strategies still exist. At this point you have to go through the best responses of the post-IESDS game as usual. However, when it does work, IESDS saves you a lot of time. As such, always pinpoint the strictly dominated strategies before you move on to weakly dominated strategies.

Takeaway Points

1) If IESDS leaves a single outcome, that outcome is the unique Nash equilibrium.
2) IESDS never eliminates Nash equilibria, but IEWDS can.
3) Always remove a strictly dominated strategy before removing a weakly dominated strategy.

Lesson 1.5: Matching Pennies and Mixed Strategy Nash Equilibrium

Here is a simple game. You and I each have a penny. Simultaneously, we choose whether to put our penny on the table with heads or tails facing up. If both of the pennies show heads or both of the pennies show tails (that is, they match), then you have to pay me a dollar. But if one shows heads and the other shows tails (that is, they do not match), then I have to pay you a dollar.

What is the optimal way to play this game? As usual, drawing a payoff matrix helps:

	Left	Right
Up	1, -1	-1, 1
Down	-1, 1	1, -1

Matching pennies is an example of a strictly competitive (or *zero sum*) game. In the prisoner's dilemma and stag hunt, the players had incentive to cooperate with each other to achieve mutually beneficial outcomes. Here, however, the players actively want to see the other perform poorly; player 1 wins whatever player 2 loses, and vice versa.

Many real world examples fit these payoffs. For example, on a soccer penalty kick, a kicker can kick left or right; the goalkeeper can dive left or right. In American football, the offense can choose a running play or passing play; the defense can choose to protect against the pass or the run. In baseball, a pitcher can throw a fastball or a curveball; a batter can guess fastball or curveball. In war, an army can invade one of two cities; the defending side can adequately defend only one of them. In each of these games, one player is happy if their strategies match, while the other player prefers a mismatch.

Although all of these alternative examples are pertinent, let's stick to the matching pennies framework. Without even marking the best responses, clearly no pure strategy Nash equilibria exist in this game. Going through the logic, suppose player 1 always played heads. Then player 2 should play tails, since that will win her a dollar. But if player 2 is playing tails as a pure strategy, player 1 should play tails as well, as that will win him the dollar. But given that player 1 is playing tails, player 2 should switch over to heads. Yet if player 2 is playing heads, player 1 should go back to playing heads. The process now begins to loop, as now player 2's best response is to play tails. We have returned to our starting point.

If we analyzed the game by marking best responses, we would wind up with this:

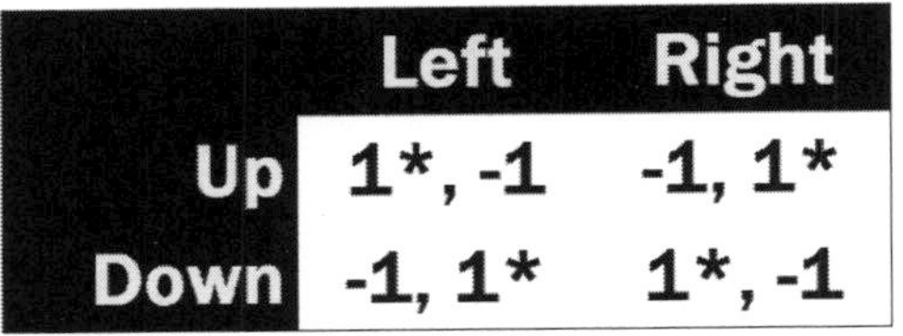

	Left	Right
Up	1*, -1	-1, 1*
Down	-1, 1*	1*, -1

Unsurprisingly, every 1 has an asterisk and every -1 does not. No outcome has an asterisk over each player's payoff, meaning no mutual best response exists in pure strategies. Therefore, the game has no pure strategy Nash equilibrium.

Does that mean this game has no solution? Not nearly. As it turns out, *every* finite game has at least one Nash equilibrium. (John Forbes Nash, of *A Beautiful Mind* fame, proved this result, which is why we call this Nash's theorem. It is also why we search for "Nash" equilibria.) A game is finite if the number of players is finite and the number of pure strategies each player has is finite. Here, there are two players, and each player has two pure strategies. Therefore, matching pennies is a finite game, so it must have a Nash equilibrium. We know matching pennies has no pure strategy Nash equilibria, so it must have a Nash equilibrium in mixed strategies.

1.5.1: What Is a Mixed Strategy?

Suppose I could read your mind. That is, as you were deciding whether to place your coin with heads or tails showing, I could perfectly anticipate what you were about to do. Is there any way you could minimize your losses?

The easiest solution is to flip the coin and cover it with your hand as it is about to land on the table. If you did, my mind reading skills would be rendered harmless. No matter what I chose, my net expected gain from playing the game would be $0. Put differently, if I guessed heads every time, I would win $1 half the time and lose $1 half the time, for an average of $0. Likewise, if I guessed tails every time, I would win $1 half the time and lose $1 half the time, again for an average of $0.

Moreover, if I flipped my coin in the same manner you do, I would still net $0. A quarter of the time they would both land on heads, and I would earn $1. Another quarter of the time they would both land on tails, and I would earn $1. In yet another quarter of the time, mine would land on heads and yours would land on tails, and I would lose $1. In the final quarter of the time, mine would land on tails and yours would land on heads, and I would again lose $1. On average, all of these cancel out, and my expected value of the game is equal to $0.

We can extend this further and say that *any* probability distribution over playing heads and tails would net me $0 on average as long as you kept flipping your coin. For example, if I played heads 1/3 of the time and tails 2/3 of the time, I would earn $0. Likewise, if I played heads 7/16 of the time and tails 9/16 of the time, I would still be stuck earning $0. Effectively, your coin flipping strategy is unbeatable and unexploitable.

From my perspective, if I cannot beat you, then I might as well flip my coin also. As we just saw, I will make $0 from this game. But interestingly, if I am flipping my coin, then you cannot beat me for the same reason I could not beat you. So when we both flip our coins, neither one of us has a profitable deviation from our strategies. Thus, this pair of coin flipping strategies is a *mixed strategy Nash equilibrium*. The "mixed strategy" part of the term refers to how we are randomizing over multiple strategies rather than playing a single "pure" strategy.

Although this coin flipping strategy is neat, it will not take us very far. Consider the following game:

	Left	Right
Up	3, -3	-2, 2
Down	-1, 1	0, 0

This game is very similar to matching pennies, except the values exchanged differ depending on the outcome; anywhere from 0 or 3 points changes hands depending on the outcome.

There are no pure strategy Nash equilibria. If player 1 is playing up, then player 2 will want to play right and earn 2. But if player 2 is playing right, then player 1 will want to play down and earn 0 instead of -2. Yet if player 1 goes down, then player 2 can deviate to left and earn 1. However, that causes player 1 to switch to up and earn 3. Now player 2 wants to go right, and we have a new cycle.

We can see this with the best responses marked:

	Left	Right
Up	3*, -3	-2, 2*
Down	-1, 1*	0*, 0

Matching pennies had the same pattern, with no outcome being a mutual best response.

Since no pure strategy Nash equilibria exist, there must be an equilibrium in mixed strategies. But is merely flipping a coin sufficient? Unfortunately, no. Suppose player 1 flipped a fair coin and played up on heads and down on tails. Then player 2's expected utility for playing left is a weighted combination of her left column payoffs. We can write it as this:

$EU_{left} = (.5)(-3) + (.5)(1) = -1$

To see how we arrived at this equation, let's break down the game a step further. We know that player 1 is playing both up and down some portion of the time, and we also know player 2 is always playing left. Since we only want to calculate player 2's payoff, we can block off all irrelevant information and end up with this:

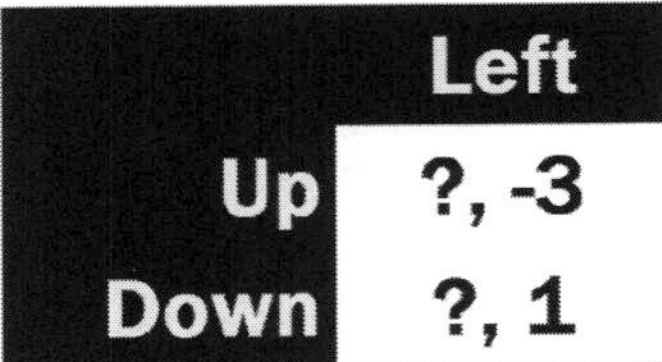

	Left
Up	?, -3
Down	?, 1

Half the time, or (.5) of the time, player 1 plays up, and player 2 earns -3. That gives us the (.5)(-3) part of the equation. The second half follows similarly. Half the time, player 1 plays down, at which point player 2 earns 1. Thus, we get the (.5)(1) part. Summing those two parts together yields the answer of -1.

Let's do the same for player 2 playing right. Blocking out the irrelevant information gives us this:

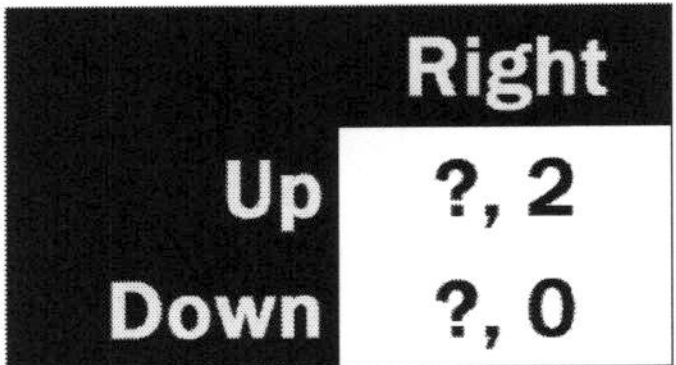

	Right
Up	?, 2
Down	?, 0

Half of the time, she earns 2. The other half of the time, she earns 0. So her expected utility of playing right in response to player 1 flipping a coin equals:

$EU_{right} = (.5)(2) + (.5)(0) = 1$

This is a problem. If player 1 flips a coin, player 2 ought to play right as pure strategy since it generates a higher expected payoff for her. Consequently, player 2 would not want to randomize between left and right in response to player 1 flipping, as doing so feeds her more of the negative payoff from left and less of the positive payoff from right.

As such, player 2's best response is to play right as a pure strategy. But if she plays right, we already know player 1's best response is to play down as a pure strategy. Now player 1 has abandoned his mixture, and we are back to where we started, as the players cycle around their pure strategy best responses.

1.5.2: The Mixed Strategy Algorithm

Fortunately, we can use a specific algorithm to find how a player can induce his or her opponent to be indifferent between the opponent's two pure strategies. Unfortunately, we now enter the computationally intensive part of game theory. Before, all we had to do was compare numbers and see which was larger. With mixed strategies, we have to introduce unknown variables and the necessary algebra to solve for them.

Let the lower case Greek letter sigma (σ) represent the probability that a player plays a particular pure strategy. For example, we use σ_{up} as the probability player 1 plays up. Using this notation, we can write player 2's expected utility of playing left as a pure strategy as a function of player 1's mixed strategy:

$$EU_{left} = (\sigma_{up})(-3) + (\sigma_{down})(1)$$

And player 2's expected utility of playing right as a pure strategy is:

$$EU_{right} = (\sigma_{up})(2) + (\sigma_{down})(0)$$

We are looking for a mixed strategy from player 1 that leaves player 2 indifferent between her pure strategies. Put differently, we want to find a σ_{up} and σ_{down} such that:

$$EU_{left} = EU_{right}$$
$$(\sigma_{up})(-3) + (\sigma_{down})(1) = (\sigma_{up})(2) + (\sigma_{down})(0)$$

Although this may seem like a difficult task, note that σ_{up} and σ_{down} must sum to one. That is, since we know player 1 only has two strategies, he must play down whenever he does not play up. Therefore, we can use the substitution $\sigma_{down} = 1 - \sigma_{up}$ to rewrite that equation:

$$(\sigma_{up})(-3) + (\sigma_{down})(1) = (\sigma_{up})(2) + (\sigma_{down})(0)$$

$\sigma_{down} = 1 - \sigma_{up}$
$(\sigma_{up})(-3) + (1 - \sigma_{up})(1) = (\sigma_{up})(2) + (1 - \sigma_{up})(0)$

And from there, all we need to do is solve for σ_{up}, and we will have our answer:

$(\sigma_{up})(-3) + (1 - \sigma_{up})(1) = (\sigma_{up})(2) + (1 - \sigma_{up})(0)$
$-3\sigma_{up} + 1 - \sigma_{up} = 2\sigma_{up}$
$1 - 4\sigma_{up} = 2\sigma_{up}$
$6\sigma_{up} = 1$
$\sigma_{up} = 1/6$

So if player 1 plays up with probability 1/6 and down with probability 5/6, player 2 earns the same payoff for selecting either left or right as a pure strategy. Consequently, she also receives the same payoff if she chooses any mixture between left and right. This is important because player 2 also has to play a mixed strategy in the Nash equilibrium of this game.

Knowing that, let's calculate the mixed strategy for player 2 that leaves player 1 indifferent between his two pure strategies. First, we need to find player 1's payoff for playing up as a function of player 2's mixed strategy σ_{left}:

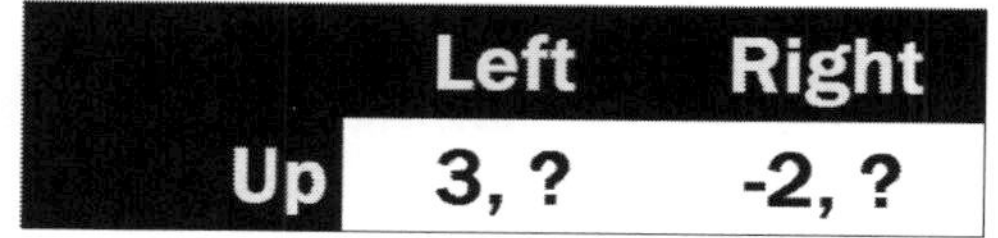

	Left	Right
Up	3, ?	-2, ?

With probability σ_{left}, player 1 earns 3; with probability $1 - \sigma_{left}$, he earns -2. We can write this as:

$EU_{up} = (\sigma_{left})(3) + (1 - \sigma_{left})(-2)$

Now we move to player 1's payoff for playing down as a function of that same mixed strategy of player 2's:

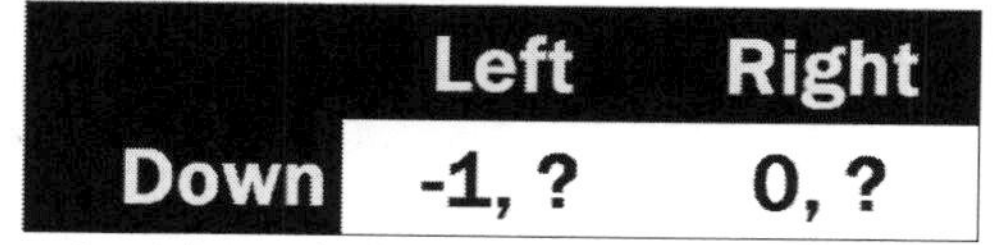

	Left	Right
Down	-1, ?	0, ?

Player 1 earns -1 with probability σ_{left} and 0 with probability $1 - \sigma_{left}$. As a formula:

$EU_{down} = (\sigma_{left})(-1) + (1 - \sigma_{left})(0)$

We need to find a mixed strategy for player 2 that leaves player 1 indifferent between his pure strategies. To do this, we set $EU_{up} = EU_{down}$ and solve for σ_{left}:

$EU_{up} = EU_{down}$
$EU_{up} = (\sigma_{left})(3) + (1 - \sigma_{left})(-2)$
$EU_{down} = (\sigma_{left})(-1) + (1 - \sigma_{left})(0)$
$(\sigma_{left})(3) + (1 - \sigma_{left})(-2) = (\sigma_{left})(-1) + (1 - \sigma_{left})(0)$
$3\sigma_{left} - 2 + 2\sigma_{left} = -\sigma_{left}$
$5\sigma_{left} - 2 = -\sigma_{left}$
$6\sigma_{left} = 2$
$\sigma_{left} = 1/3$

So if player 2 plays left with probability 1/3 and right with probability 2/3, player 1 is indifferent between playing up and down as pure strategies. Moreover, he is indifferent between this and playing any mixture between up and down, as that mixture will still provide him the same payoff.

Connecting the mixed strategies of both players together, we see they are a best response to each other and therefore a Nash equilibrium. That is, if player 1 plays up with probability 1/6 and down with probability 5/6, then any strategy of player 2's produces the same payoff, so any mixture between her two strategies is a best response. This includes mixing left with probability 1/3 and down with probability 2/3. Likewise, if player 2 plays left with probability 1/3 and right with probability 2/3, any strategy of player 1's generates the same payoff, so any mixture between his two strategies is a best response. This includes mixing up with probability 1/6 and down with probability 5/6. Since neither player can profitably change his or her strategy, those mixtures are a mixed strategy Nash equilibrium.

1.5.3: How NOT to Write a Mixed Strategy Nash Equilibrium

In the previous game, player 2 played left with probability 1/3 and right with probability 2/3. If we solved for her probability of playing left with a calculator, we would see something along the lines of 0.33333333. Likewise, the calculator would generate 0.66666666 for her probability of playing right. Many students substitute 0.33 and 0.67 for 1/3 and 2/3. But are 0.33 and 0.67 technically correct?

The answer is no. To see why, suppose player 2 played left with probability 0.33 and right with probability 0.67. For this to be a mixed strategy Nash equilibrium, player 1 must be indifferent between playing up and down. Knowing that, let's calculate player 1's expected utility of playing up and his expected utility of playing down. We will begin with up:

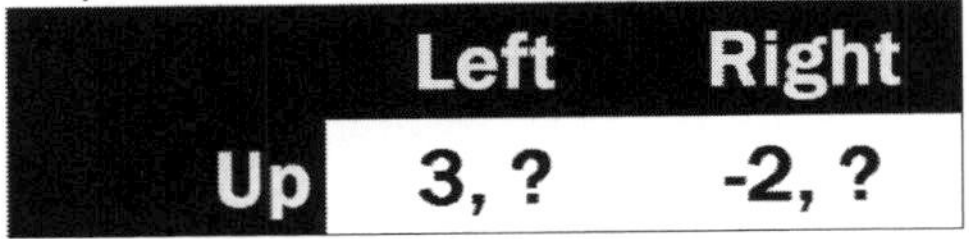

	Left	Right
Up	3, ?	-2, ?

If he plays up, he earns 3 with probability 0.33 and -2 with probability 0.67. As an equation:

$EU_{up} = (0.33)(3) + (0.67)(-2)$
$EU_{up} = 0.99 - 1.34$
$EU_{up} = -0.35$

Now let's find his expected utility if he plays down:

	Left	Right
Down	-1, ?	0, ?

This time, he earns -1 with probability 0.33 and 0 with probability 0.67. As an equation:

$EU_{down} = (0.33)(-1) + (0.67)(0)$
$EU_{down} = -0.33$

It should be apparent why this is not a Nash equilibrium. If player 2 actually played left with probability 0.33 and right with probability 0.67, player 1's best response is to play down as a pure strategy, as -0.33 is greater than -0.35. The margin is slight, but down is better nonetheless. To induce indifference, the probabilities must be *exactly* 1/3 and *exactly* 2/3. Although 0.33 may seem innocuously different from 1/3, anything even slightly more or slightly less than 1/3 breaks this all-important indifference.

Fortunately, there is a simple way to avoid this problem: write all of your answers as fractions. Decimals lose the precision necessary for game theoretical calculations, but fractions always work perfectly. If you get into the habit of using fractions exclusively, you will not lose points on an exam on such a technicality.

1.5.4: Mixed Strategies in the Stag Hunt

Some games have both pure strategy Nash equilibria and mixed strategy Nash equilibria. The stag hunt from two lessons ago is one such game. Recall that both <stag, stag> and <hare, hare> are PSNE. However, a

MSNE also exists in which both players mix between stag and hare. We can use the same algorithm from before to solve for it.

To differentiate between the players' strategies and for ease of exposition, let's replace the names of the players' strategies with up, down, left, and right to give us this game:

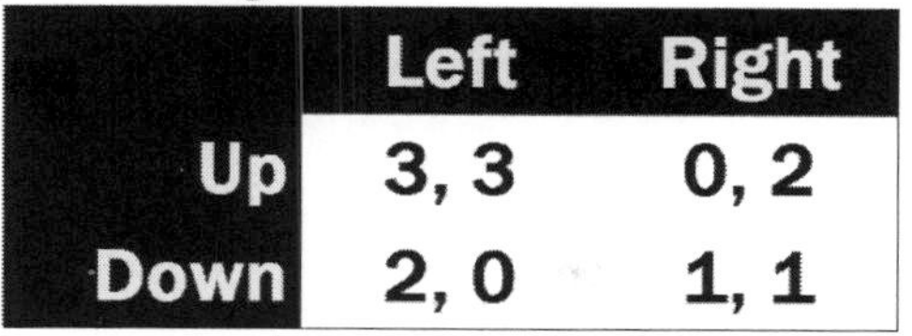

	Left	Right
Up	3, 3	0, 2
Down	2, 0	1, 1

We start by finding player 2's expected utility for playing left as a pure strategy as a function of player 1's mixed strategy σ_{up}:

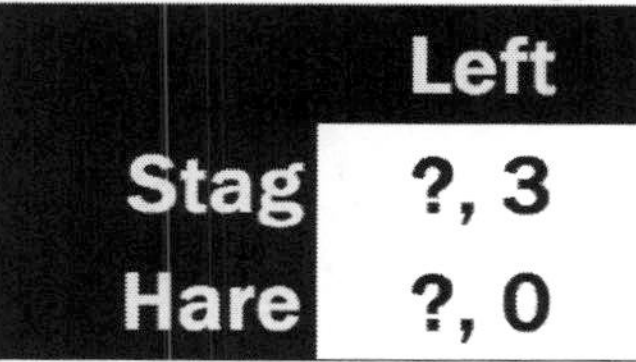

	Left
Stag	?, 3
Hare	?, 0

So with probability σ_{up}, player 2 earns 3. With complementary probability, she earns 0. As an equation:

$EU_{left} = (\sigma_{up})(3) + (1 - \sigma_{up})(0)$

Now we go to player 2's expected utility for playing right as a pure strategy as a function of that same mixed strategy σ_{up}:

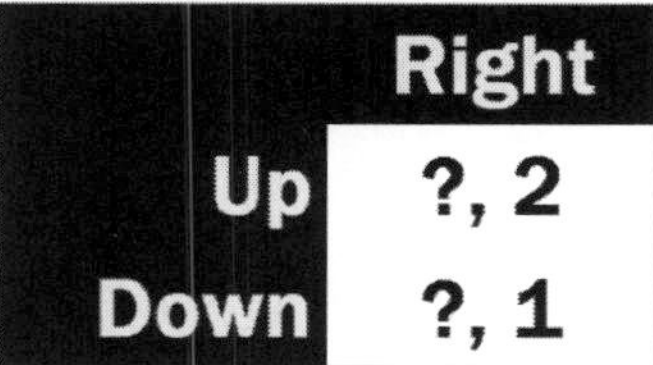

	Right
Up	?, 2
Down	?, 1

Here, player 2 earns 2 with probability σ_{up} and 1 with probability $1 - \sigma_{up}$. Again, to write that as an equation:

$EU_{right} = (\sigma_{up})(2) + (1 - \sigma_{up})(1)$

To induce player 2's indifference, we set her expected utility for playing left equal to her expected utility for playing right and solve for σ_{up}:

$EU_{left} = EU_{right}$
$EU_{left} = (\sigma_{up})(3) + (1 - \sigma_{up})(0)$
$EU_{right} = (\sigma_{up})(2) + (1 - \sigma_{up})(1)$

$(\sigma_{up})(3) + (1 - \sigma_{up})(0) = (\sigma_{up})(2) + (1 - \sigma_{up})(1)$
$3\sigma_{up} = 2\sigma_{up} + 1 - \sigma_{up}$
$3\sigma_{up} = \sigma_{up} + 1$
$2\sigma_{up} = 1$
$\sigma_{up} = 1/2$

So if player 1 plays up and down each with probability 1/2, player 2 is indifferent between all combinations of left and right.

To complete the MSNE, we have to see what mixed strategy for player 2 leaves player 1 indifferent between his pure strategies. The process is the same, so we begin by looking at player 1's expected utility for playing up as a pure strategy as a function of player 2's mixed strategy σ_{left}:

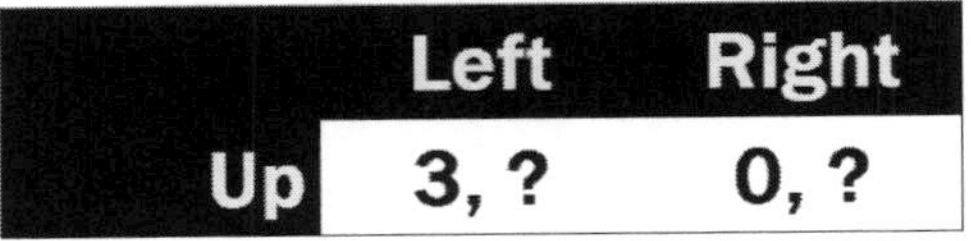

	Left	Right
Up	3, ?	0, ?

He earns 3 with probability σ_{left} and 0 with probability $1 - \sigma_{left}$, or:

$EU_{up} = (\sigma_{left})(3) + (1 - \sigma_{left})(0)$

Now for down:

	Left	Right
Down	2, ?	1, ?

Here, he earns 2 with probability σ_{left} and 1 with probability $1 - \sigma_{left}$, or:

$EU_{down} = (\sigma_{left})(2) + (1 - \sigma_{left})(1)$

Combining these together yields the following:

$EU_{up} = EU_{down}$
$EU_{up} = (\sigma_{left})(3) + (1 - \sigma_{left})(0)$
$EU_{down} = (\sigma_{left})(2) + (1 - \sigma_{left})(1)$
$(\sigma_{left})(3) + (1 - \sigma_{left})(0) = (\sigma_{left})(2) + (1 - \sigma_{left})(1)$
$3\sigma_{left} = 2\sigma_{left} + 1 - \sigma_{left}$
$3\sigma_{left} = \sigma_{left} + 1$
$2\sigma_{left} = 1$
$\sigma_{left} = 1/2$

Thus, player 2 plays left and right each with probability 1/2 in the MSNE.

1.5.5: How Changing Payoffs Affects Mixed Strategy Nash Equilibria

When we solved the prisoner's dilemma, we saw that changing the payoffs did not matter much. Indeed, as long as the order of the numbers stayed the same—that is, we replaced the largest number from the first set of payoffs with the largest number from the second set of payoffs, we replaced the second largest number from the first set of payoffs with the second largest number from the second set of payoffs, and so forth—the equilibrium did not change. Regardless of the individual values, as long as the order of the numbers remained the same, confessing still strictly dominated keeping quiet, and thus <confess, confess> was the unique pure strategy Nash equilibrium of the game.

Mixed strategy Nash equilibrium is shakier. Although the MSNE will still exist as long as the order remains the same, even slight perturbations in relative values alters the exact probability distributions found in the MSNE.

To see this in action, let's look at a slightly altered stag hunt game:

	Left	Right
Up	4, 4	0, 2
Down	2, 0	1, 1

Let's solve for player 1's mixed strategy. As before, this requires us to define player 2's expected utility for each of her pure strategies as a function of player 1's mixed strategy σ_{up}. Let's begin with her expected utility for left as a pure strategy:

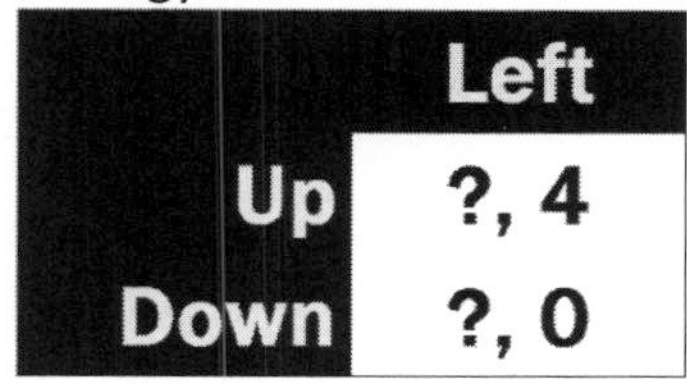

	Left
Up	?, 4
Down	?, 0

Player 2 receives 4 with probability σ_{up} and 0 with probability $1 - \sigma_{up}$. Therefore, her expected utility is:

$$EU_{left} = (\sigma_{up})(4) + (1 - \sigma_{up})(0)$$

And now for player 2's expected utility for right:

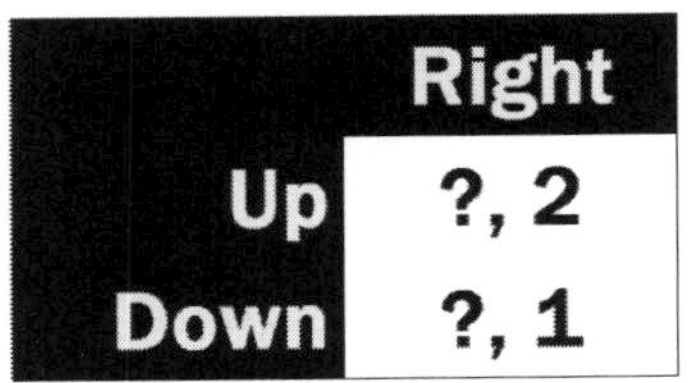

	Right
Up	?, 2
Down	?, 1

This time, she earns 2 with probability σ_{up} and 1 with probability $1 - \sigma_{up}$. We can write this as:

$EU_{right} = (\sigma_{up})(2) + (1 - \sigma_{up})(1)$

Note that the expected utility for playing right remained the same, but the expected utility for left changed. As such, when we go through the last step of the process (setting EU_{left} equal to EU_{right} and solving for σ_{up}), we should expect a different mixed strategy. Sure enough, this is true:

$EU_{left} = EU_{right}$
$EU_{left} = (\sigma_{up})(4) + (1 - \sigma_{up})(0)$
$EU_{right} = (\sigma_{up})(2) + (1 - \sigma_{up})(1)$
$(\sigma_{up})(4) + (1 - \sigma_{up})(0) = (\sigma_{up})(2) + (1 - \sigma_{up})(1)$
$4\sigma_{up} = 2\sigma_{up} + 1 - \sigma_{up}$
$4\sigma_{up} = \sigma_{up} + 1$
$3\sigma_{up} = 1$
$\sigma_{up} = 1/3$

So player 1 plays up with probability 1/3 and down with probability 2/3 in the MSNE. The same steps will also show that player 2 makes the same change, now mixing left with probability 1/3 and right with probability 2/3. These mixtures are different from when the <stag, stag> outcome was worth 3 for each player. Consequently, we must recalculate mixed strategy Nash equilibria every time a payoff changes.

1.5.6: Invalid Mixed Strategies

Not all games have a mixed strategy Nash equilibrium. Deadlock, for example, does not. To see this, let's try to use the algorithm to solve for an imaginary MSNE of the game. Recall that deadlock looked like this:

	Left	Right
Up	3, 3	4, 1
Down	1, 4	2, 2

Let's find the mixed strategy for player 2 that leaves player 1 indifferent between his pure strategies. We begin by isolating player 1's payoffs if he plays up as a pure strategy:

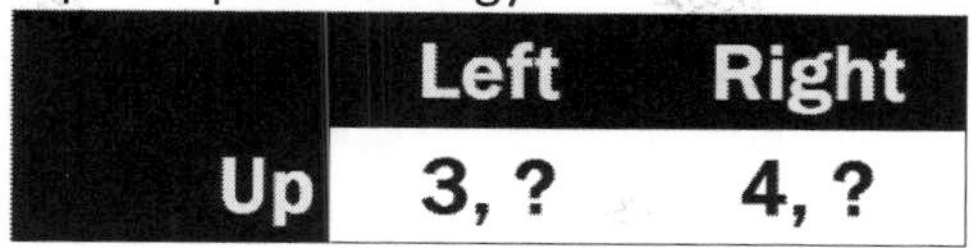

	Left	Right
Up	3, ?	4, ?

If player 2 plays σ_{left} as her mixed strategy and player 1 plays up as a pure strategy, he earns 3 with probability σ_{left} and 4 with probability $1 - \sigma_{left}$. As an equation:

$EU_{up} = (\sigma_{left})(3) + (1 - \sigma_{left})(4)$

Now consider player 1 selecting down as a pure strategy:

	Left	Right
Down	1, ?	2, ?

This time, he earns 1 with probability σ_{left} and 2 with probability $1 - \sigma_{left}$. We write this as:

$EU_{down} = (\sigma_{left})(1) + (1 - \sigma_{left})(2)$

Let's attempt to set EU_{up} equal to EU_{down} and solve for σ_{left}:

$EU_{up} = EU_{down}$
$EU_{up} = (\sigma_{left})(3) + (1 - \sigma_{left})(4)$
$EU_{down} = (\sigma_{left})(1) + (1 - \sigma_{left})(2)$
$(\sigma_{left})(3) + (1 - \sigma_{left})(4) = (\sigma_{left})(1) + (1 - \sigma_{left})(2)$
$3\sigma_{left} + 4 - 4\sigma_{left} = \sigma_{left} + 2 - 2\sigma_{left}$
$-\sigma_{left} + 4 = -\sigma_{left} + 2$
$2 = 0$

In the process of solving for the mixed strategy, we eliminated σ_{left} from the equation in the last step and made the bold mathematical claim that 2 = 0. This is an absurd statement. Essentially, our algorithm is telling us that no mixed strategy that can make this work. Thus, we can safely conclude that player 2 cannot mix in a way that leaves player 1 indifferent. The same steps would also show that player 1 cannot mix in a manner that leaves player 2 indifferent. As a result, no MSNE exists.

Other things can go wrong when we try running the mixed strategy algorithm. Recall that the original prisoner's dilemma had the following payoff matrix:

	Left	Right
Up	-1, -1	-12, 0
Down	0, -12	-8, -8

Let's solve for player 1's mixed strategy. We begin by isolating player 2's payoffs for playing left (quiet):

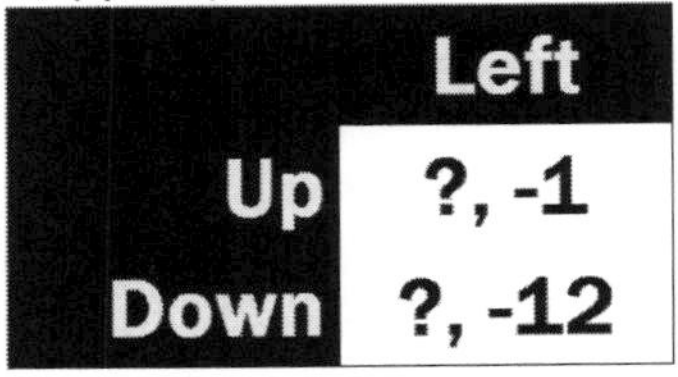

	Left
Up	?, -1
Down	?, -12

If player 2 selects left as a pure strategy, she earns -1 with probability σ_{up} and -12 with probability $1 - \sigma_{up}$. Thus, her expected utility equals:

$EU_{left} = (\sigma_{up})(-1) + (1 - \sigma_{up})(-12)$

Now isolate player 2's payoffs for playing right (confess):

	Right
Up	?, 0
Down	?, -8

Here, she earns 0 with probability σ_{up} and -8 with probability $1 - \sigma_{up}$. So her expected utility equals:

$EU_{right} = (\sigma_{up})(0) + (1 - \sigma_{up})(-8)$

Let's try to solve for the mixed strategy that induces indifference:

$EU_{left} = EU_{right}$
$EU_{left} = (\sigma_{up})(-1) + (1 - \sigma_{up})(-12)$
$EU_{right} = (\sigma_{up})(0) + (1 - \sigma_{up})(-8)$
$(\sigma_{up})(-1) + (1 - \sigma_{up})(-12) = (\sigma_{up})(0) + (1 - \sigma_{up})(-8)$
$-\sigma_{up} - 12 + 12\sigma_{up} = -8 + 8\sigma_{up}$
$-12 + 11\sigma_{up} = -8 + 8\sigma_{up}$
$3\sigma_{up} = 4$

$\sigma_{up} = 4/3$

Player 1 is supposed to play up with probability 4/3. We know that the probability he plays up is therefore 1 – 4/3, or -1/3. As such, he plays up with a probability greater than 100% and down a negative percentage of the time. Neither one of these things is sensible; all probabilities must fall between 0 and 1. Once more, this is our algorithm's way of telling us that no such MSNE exists.

These two examples demonstrate when players cannot make their opponents indifferent. Unfortunately, if we show that one player cannot mix in such a manner, we still cannot eliminate the possibility that no MSNE exists. In particular, there are games where one player plays a pure strategy while the other mixes. Some differentiate these as "partially mixed strategy Nash equilibria" because one player mixes and the other does not. Regardless, they still fall under the umbrella of MSNE. We will look at such games in Lesson 1.8.

1.5.7: Mixing and Dominance

Both the prisoner's dilemma and deadlock have a strictly dominated strategy. In addition, neither has a mixed strategy Nash equilibrium. Is this a coincidence?

Most certainly not: a strictly dominated strategy cannot be played with positive probability in a MSNE. Thinking about how mixed strategy Nash equilibria work reveals why. A requirement of Nash equilibrium is that no player has a profitable deviation from the strategies being played. So imagine that a player used a strictly dominated strategy as a part of a mixed strategy. Then some percentage of the time he ends up in an outcome which is strictly worse than if he had used the strictly dominating strategy as a part of the mixed strategy instead. That implies that he can improve his payoff by playing the strictly dominating strategy whatever percentage of the time he would otherwise play the strictly dominated strategy. Thus, he has a profitable deviation whenever he plays the strictly dominated strategy.

To see how this operates, let's use the prisoner's dilemma as an example. Suppose player 1 could mix in equilibrium. Then some percentage of the time, he keeps quiet. If player 2 plays keep quiet as a pure strategy, player 1 would regret playing keep quiet whenever his mixture told him to do so, as he could have played confess and spent less time in jail. Likewise, if player 2 plays confess as a pure strategy, player 1 would again regret playing keep quiet whenever his mixture told him to do so for the same reason.

The logic is a little more complicated if player 2 were mixing. Suppose player 1 gets out his randomizing device, and it tells him to keep quiet. Given that, he knows player 2 confesses with probability $\sigma_{confess}$ and keeps quiet with probability $1 - \sigma_{confess}$, but he still does not know which until he actually reveals his choice to her. However, regardless of what she ends up selecting, player 1 could perform better by confessing. That is, for the $\sigma_{confess}$ amount of time, he would rather confess, and for the $1 - \sigma_{confess}$, he would still rather confess. The bottom line is that it makes no sense for him to play the strictly dominated strategy in a mixed strategy, so he does not.

Weakly dominated strategies remain finicky, however. They can be played in a MSNE. We will see such a case in Lesson 1.8.

Takeaway Points

1) Nash's theorem states that every finite game has at least one Nash equilibrium.
2) The mixed strategy algorithm derives mixed strategy Nash equilibria by finding the particular mixed strategies that leave the other player indifferent between his or her two pure strategies.
3) Expressing mixed strategies in fractions is more accurate than expressing them in decimals.
4) Players cannot mix using strictly dominated strategies in equilibrium.

Lesson 1.6: Calculating Payoffs

When players play pure strategy Nash equilibria, their payoffs are obvious. However, in a mixed strategy Nash equilibrium, we cannot simply look at an outcome and read a number to find a player's expected utility. This lesson shows how to calculate payoffs in MSNE using three commonly seen games: chicken, battle of the sexes, and pure coordination.

1.6.1: Chicken

Two testosterone-fueled teenagers are on opposite ends of a one lane street and begin driving full speed toward one another. At the last possible moment, they must decide whether to swerve or continue going straight. If one continues while the other swerves, the one who swerves is a "chicken" while the other has proven his or her bravery. If both swerve, then neither can claim superiority. But if both continue, they crash straight into each other in an epic conflagration, which is the worst possible outcome for both players.

Although these teenagers' preferences may seem strange to us, we can still analyze their optimal behavior (from their perspectives) using the following utilities over outcomes:

	Continue	Swerve
Continue	-10, -10	2, -2
Swerve	-2, 2	0, 0

"Snowdrift" is an alternative title for this game. Under that framework, two drivers are stuck on the opposite ends of a snowy road, and they simultaneously decide whether to stay in their cars or shovel a passageway. An individual's most preferred outcome is to sit in his or her car while the other player shovels. That individual's next best outcome is for both players to shovel. The third best outcome is for that player to do the shoveling while the other stays inside. The worst outcome is for both players to stay in their cars, as the road never clears.

We will stick to chicken because it allows for the possibility of a fiery explosion and does not involve any depressing wintery weather. This game has two pure strategy Nash equilibria: <continue, swerve> and <swerve, continue>. Let's verify this by marking each player's best responses, beginning with how player 1 should respond to player 2 continuing:

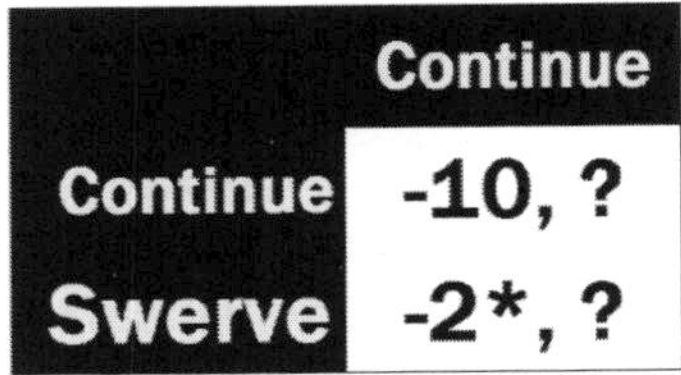

	Continue
Continue	-10, ?
Swerve	-2*, ?

Although both outcomes are undesirable to player 1, swerving minimizes his losses, so it earns the asterisk.

Next, suppose player 2 swerved:

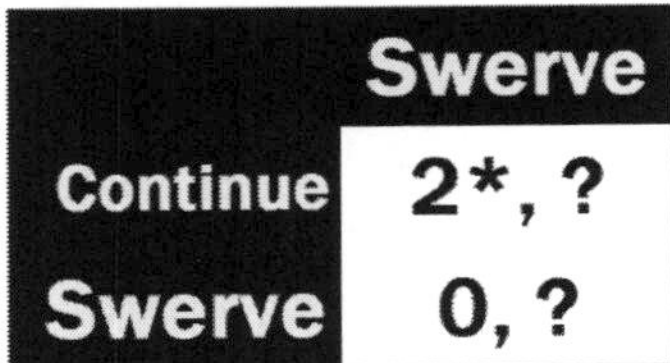

	Swerve
Continue	2*, ?
Swerve	0, ?

Player 2 swerving opens up better outcomes for player 1. However, continuing is better than swerving here, meaning that player 1 wants to condition his behavior on how player 2 acts.

The same is true for player 2. First, suppose player 1 continued:

	Continue	Swerve
Continue	?, -10	?, -2*

Then player 2 wants to swerve. But if he swerves:

	Continue	Swerve
Swerve	?, 2*	?, 0

Then she wants to continue. Hence, this is a game of mixed motives. Both players want to continue and force the other player to swerve. But if neither backs down, the game ends in disaster. If one can convince the other to back down, the game reaches a Nash equilibrium:

	Continue	Swerve
Continue	-10, -10	2*, -2*
Swerve	-2*, 2*	0, 0

That is, <continue, swerve> and <swerve, continue> are pure strategy Nash equilibria since both players' payoffs have stars in those boxes.

However, each player cannot credibly threaten to continue if he or she knows the other will continue as well, as that leads to the disastrous <continue, continue> outcome. So perhaps a MSNE exists as well. To find out, let's begin by renaming the strategies as directions to distinguish between each player's moves:

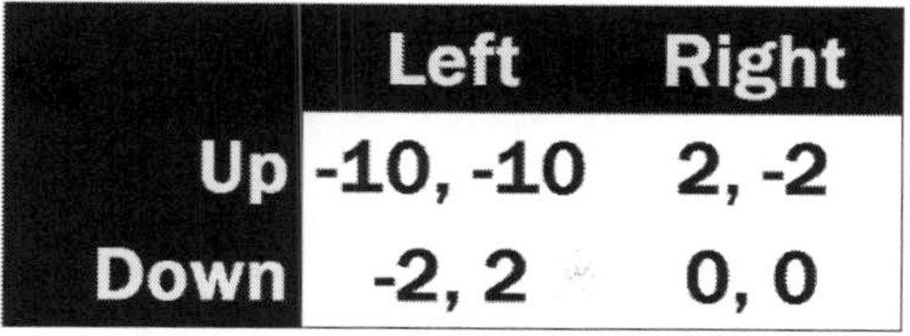

	Left	Right
Up	-10, -10	2, -2
Down	-2, 2	0, 0

First, let's solve for player 1's mixed strategy. Recall that we need to find player 2's payoffs for each of her pure strategies as a function of player 1's mixture.

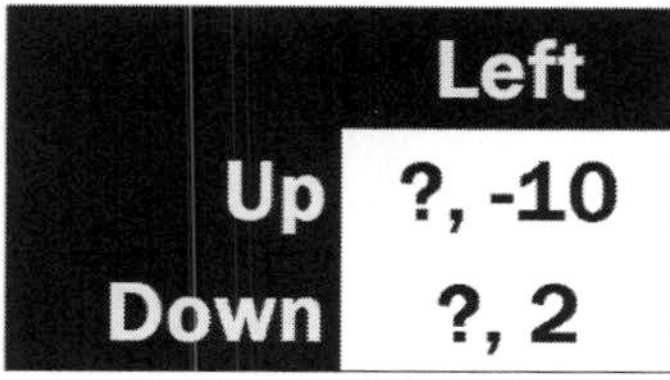

	Left
Up	?, -10
Down	?, 2

Player 2 earns -10 with probability σ_{up} and 2 with probability $1 - \sigma_{up}$. Therefore, her expected utility for left equals:

$EU_{left} = (\sigma_{up})(-10) + (1 - \sigma_{up})(2)$

Now for right:

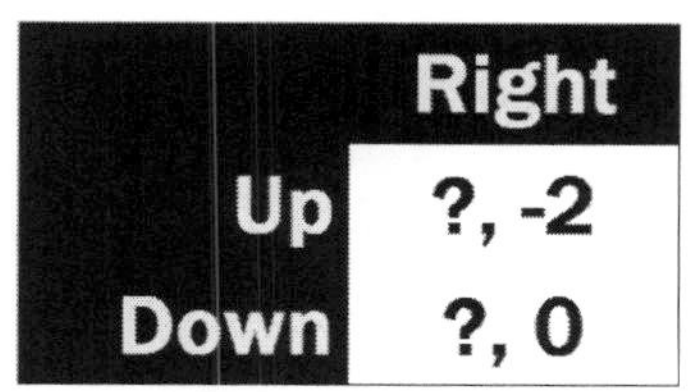

	Right
Up	?, -2
Down	?, 0

Here, player 2 earns -2 with probability σ_{up} and 0 with probability $1 - \sigma_{up}$, or:

$EU_{right} = (\sigma_{up})(-2) + (1 - \sigma_{up})(0)$

Setting these two expected utilities equal to each other allows us to solve for player 1's mixed strategy that leaves player 2 indifferent:

$EU_{left} = EU_{right}$
$EU_{left} = (\sigma_{up})(-10) + (1 - \sigma_{up})(2)$
$EU_{right} = (\sigma_{up})(-2) + (1 - \sigma_{up})(0)$
$(\sigma_{up})(-10) + (1 - \sigma_{up})(2) = (\sigma_{up})(-2) + (1 - \sigma_{up})(0)$
$-10\sigma_{up} + 2 - 2\sigma_{up} = -2\sigma_{up}$
$-12\sigma_{up} + 2 = -2\sigma_{up}$

$10\sigma_{up} = 2$
$\sigma_{up} = 1/5$

So player 1 goes up (continues) with probability 1/5 and goes down (swerves) with probability 4/5.

Using the same process for player 2's mixed strategy, we find an identical result. We begin with up as a pure strategy for player 1:

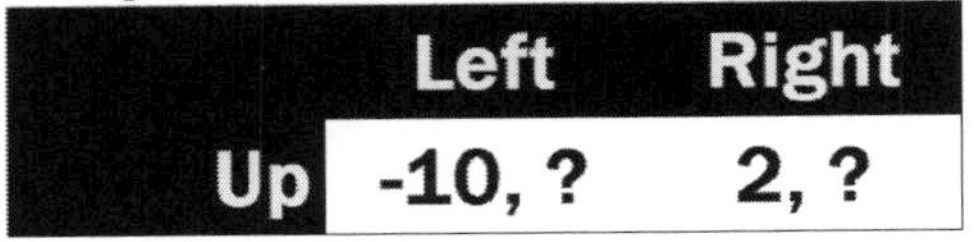

	Left	Right
Up	-10, ?	2, ?

So with probability σ_{left}, player 1 earns -10, and with probability $1 - \sigma_{left}$, he earns 2. As an equation:

$EU_{up} = (\sigma_{left})(-10) + (1 - \sigma_{left})(2)$

For down:

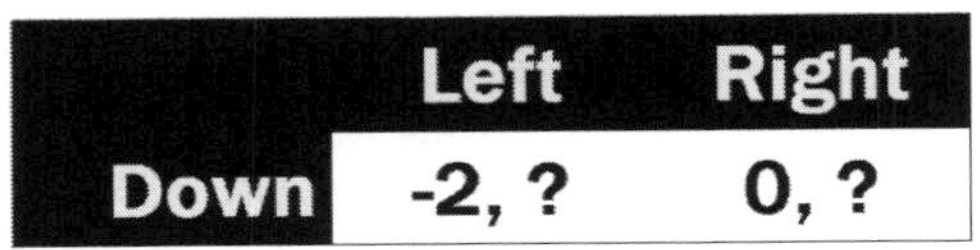

	Left	Right
Down	-2, ?	0, ?

Now he receives -2 with probability σ_{left} and 0 with probability $1 - \sigma_{left}$. We can write this as:

$EU_{down} = (\sigma_{left})(-2) + (1 - \sigma_{left})(0)$

Setting player 1's expected utility for up equal to his expected utility for down and solving for σ_{left} yields:

$EU_{up} = EU_{down}$
$EU_{up} = (\sigma_{left})(-10) + (1 - \sigma_{left})(2)$
$EU_{down} = (\sigma_{left})(-2) + (1 - \sigma_{left})(0)$
$(\sigma_{left})(-10) + (1 - \sigma_{left})(2) = (\sigma_{left})(-2) + (1 - \sigma_{left})(0)$
$-10\sigma_{left} + 2 - 2\sigma_{left} = -2\sigma_{left}$
$-12\sigma_{left} + 2 = -2\sigma_{left}$
$10\sigma_{left} = 2$
$\sigma_{left} = 1/5$

So in the MSNE, both players continue to drive straight with probability 1/5 and swerve with probability 4/5.

A natural question to ask is how the expected outcome of the MSNE compares to the outcomes in the PSNE. We can readily identify the value of the PSNE outcomes, since the payoff matrix explicitly tells us that the player who swerves gets -2 and the player who goes straight gets 2. However, all the outcomes occur with positive probability in the MSNE, which complicates matters. Indeed, we have to manually calculate these payoffs.

Fortunately, the process is easy. We simply take the probability that each outcome occurs in equilibrium and multiply it by the player's expected utility for that outcome. Let's isolate all of player 1's payoffs:

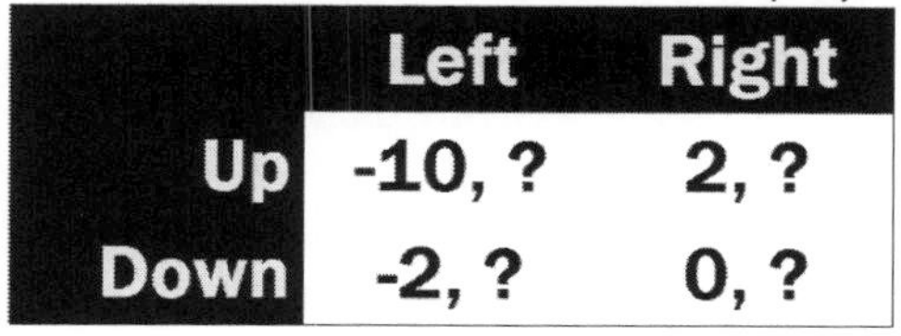

	Left	Right
Up	-10, ?	2, ?
Down	-2, ?	0, ?

We know that player 1 plays up with probability 1/5 and player 2 plays left with probability 1/5. So the probability player 1 earns his <continue, continue> payoff of -10 is simply the product of 1/5 and 1/5. Likewise, since he plays down with probability 4/5, the probability he earns his <swerve, continue> payoff of -2 is the product of 1/5 and 4/5. The probability <swerve, continue> occurs in the MSNE is the product of 4/5 and 1/5, so he earns 2 times that probability. Finally, the probability he earns 0 from the <swerve, swerve> outcome is the product of 4/5 and 4/5.

Player 1's expected utility is simply the sum of all of these products times the payoffs. In an equation form, solving for player 1's expected utility is routine:

$$EU_1 = (1/5)(1/5)(-10) + (4/5)(1/5)(-2) + (1/5)(4/5)(2) + (4/5)(4/5)(0)$$
$$EU_1 = (1/25)(-10) + (4/25)(-2) + (4/25)(2) + (16/25)(0)$$
$$EU_1 = -10/25 - 8/25 + 8/25 + 0$$
$$EU_1 = -10/25$$
$$EU_1 = -2/5$$

So in the MSNE, player 1 expects to earn -2/5 on average. Note that he will not actually earn -2/5 when they play—he will either earn -10, -2, 0, or 2. But his *average* outcome is -2/5. This is similar to what you expect when you play a game like blackjack at a casino. You *expect* to lose a few cents per dollar you spend at the table, but on any given hand you either lose your entire bet or double it (or more if you hit a blackjack). Although no single outcome exists that ends with you receiving $0.98 or so of your original $1, it is nevertheless correct to say that $0.98 is your average outcome. When we find expected utilities of MSNE, we define the expectations analogously.

Player 2 has the same expected utility. We compute this the same way, beginning by isolating her payoffs:

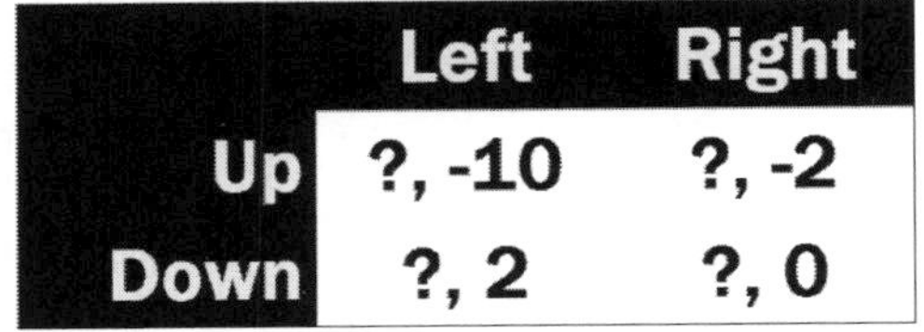

	Left	Right
Up	?, -10	?, -2
Down	?, 2	?, 0

The probability each outcome occurs in equilibrium is the same as before. As such, calculating the expected utility is just a matter of compiling these payoffs and their probabilities and then summing them together:

$EU_2 = (1/5)(1/5)(-10) + (1/5)(4/5)(2) + (4/5)(1/5)(-2) + (4/5)(4/5)(0)$
$EU_2 = (1/25)(-10) + (4/25)(2) + (4/25)(-2) + (16/25)(0)$
$EU_2 = -10/25 + 8/25 - 8/25 + 0$
$EU_2 = -10/25$
$EU_2 = -2/5$

As claimed, player 2's payoff is -2/5, just like player 1's.

Now that we know each player's expected utility in the MSNE, we can see the <swerve, swerve> outcome leaves both players better off, as 0 is greater than -2/5. However, as we saw with the prisoner's dilemma, such an outcome is inherently unstable, as one of the players could profitably deviate to continuing.

1.6.2: Battle of the Sexes

A man and a woman want to go on a date on a Friday evening in the 1980s. There are only two venues of entertainment in the city that night: a ballet and a fight. The woman wants to see the ballet. The man wants to see the fight. However, they both prefer being together than being alone, as they will have to go home immediately if the other does not show up at the location they choose. A simple cell phone call, text message, or email would simplify the coordination process, but the 1980s lacked those luxuries. As such, both must choose where to go simultaneously and without the ability to communicate with one another.

An alternative framework for this game pits two people deciding whether to go to a Johann Sebastian Bach or Igor Stravinsky concert, but the central concepts and payoffs are the same. We can draw up the payoff matrix like this:

	Ballet	Fight
Ballet	1, 2	0, 0
Fight	0, 0	2, 1

Marking best responses to check for pure strategy Nash equilibrium is overkill here. If both go to the ballet, then neither has a profitable deviation, as they would become separated and earn 0 rather than some positive amount. So <ballet, ballet> is a Nash equilibrium. Likewise, if both go to the fight, then any deviation moves a player's payoff from a positive value to 0, so <fight, fight> is also a Nash equilibrium. Meanwhile, <ballet, fight> and <fight, ballet> are not Nash equilibria, as each player would rather go to wherever his or her date is than stick to his or her current strategy.

If we check for MSNE, we see one exists for this game. As always, to make things convenient, let's rename the strategies according to directions:

	Left	Right
Up	1, 2	0, 0
Down	0, 0	2, 1

We will solve for player 1's mixed strategy first, meaning we need isolate player 2's payoffs if she moves left:

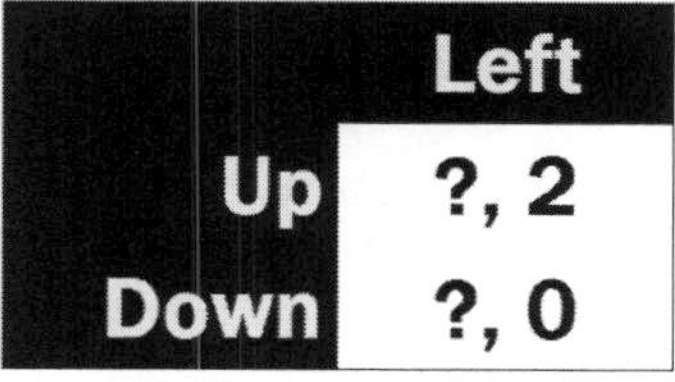

	Left
Up	?, 2
Down	?, 0

So she earns 2 with probability σ_{up} and 0 with probability $1 - \sigma_{up}$. As an equation:

$EU_{left} = (\sigma_{up})(2) + (1 - \sigma_{up})(0)$

Now consider her expected utility for playing right as a pure strategy:

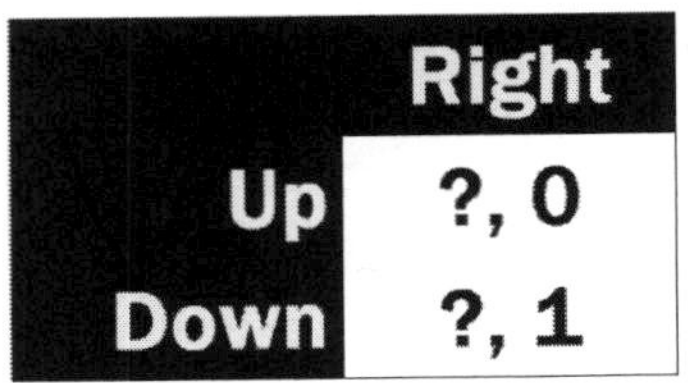

	Right
Up	?, 0
Down	?, 1

Here, she earns 0 with probability σ_{up} and 1 with probability $1 - \sigma_{up}$. Once again, as an equation:

$EU_{right} = (\sigma_{up})(0) + (1 - \sigma_{up})(1)$

To find player 1's mixed strategy that makes player 2 indifferent, we set her expected utility for left equal to her expected utility for right and solve for σ_{up}:

$EU_{left} = EU_{right}$
$EU_{left} = (\sigma_{up})(2) + (1 - \sigma_{up})(0)$
$EU_{right} = (\sigma_{up})(0) + (1 - \sigma_{up})(1)$
$(\sigma_{up})(2) + (1 - \sigma_{up})(0) = (\sigma_{up})(0) + (1 - \sigma_{up})(1)$
$2\sigma_{up} = 1 - \sigma_{up}$
$3\sigma_{up} = 1$
$\sigma_{up} = 1/3$

So in the MSNE, player 1 goes to the ballet (up) with probability 1/3 and to the fight (down) with probability 2/3.

Switching gears, let's solve for player 2's mixed strategy. To begin, consider player 1's payoffs for playing up as a function of that mixed strategy:

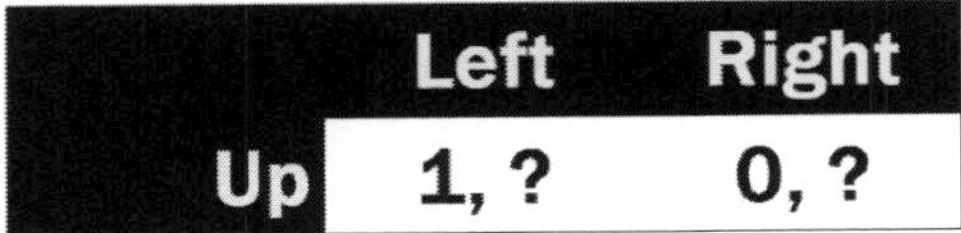

	Left	Right
Up	1, ?	0, ?

From this, we see that player 1 earns 1 with probability σ_{left} and 0 with probability $1 - \sigma_{left}$. So his expected utility for moving up equals:

$EU_{up} = (\sigma_{left})(1) + (1 - \sigma_{left})(0)$

Now consider player 1's payoffs for down:

	Left	Right
Down	0, ?	2, ?

This time it is 0 with probability σ_{left} and 2 with probability $1 - \sigma_{left}$, or:

$$EU_{down} = (\sigma_{left})(0) + (1 - \sigma_{left})(2)$$

Finally, we set his expected utility for up equal to his expected utility for down and solve for σ_{left}:

$$EU_{up} = EU_{down}$$
$$EU_{up} = (\sigma_{left})(1) + (1 - \sigma_{left})(0)$$
$$EU_{down} = (\sigma_{left})(0) + (1 - \sigma_{left})(2)$$
$$(\sigma_{left})(1) + (1 - \sigma_{left})(0) = (\sigma_{left})(0) + (1 - \sigma_{left})(2)$$
$$\sigma_{left} = 2 - 2\sigma_{left}$$
$$3\sigma_{left} = 2$$
$$\sigma_{left} = 2/3$$

So player 2 goes to the ballet (left) with probability 2/3 and the fight (right) with probability 1/3.

In summary, each player goes to his or her preferred form of entertainment with probability 2/3 and his or her lesser preferred form of entertainment with probability 1/3. But these probabilities do not tell us about the efficiency of the MSNE. To find out, we must calculate their expected utilities in the MSNE.

Let's isolate player 1's payoffs for each outcome:

	Left	Right
Up	1, ?	0, ?
Down	0, ?	2, ?

The <up, right> outcome occurs with probability 1/3 times 2/3. The <down, left> outcome occurs with probability 2/3 times 2/3. The <up, right> outcome occurs with probability 1/3 times 1/3. Finally, the <down, right> outcome occurs with probability 2/3 times 1/3. Multiplying these probabilities by their respective payoffs and then summing all of those payoffs together gives us the following:

$$EU_1 = (1/3)(2/3)(1) + (2/3)(2/3)(0) + (1/3)(1/3)(0) + (2/3)(1/3)(2)$$
$$EU_1 = (2/9)(1) + (4/9)(0) + (1/9)(0) + (2/9)(2)$$
$$EU_1 = 2/9 + 4/9$$
$$EU_1 = 6/9$$
$$EU_1 = 2/3$$

So player 1 earns a lowly 2/3 in the MSNE.

The same process reveals that player 2 earns 2/3 as well:

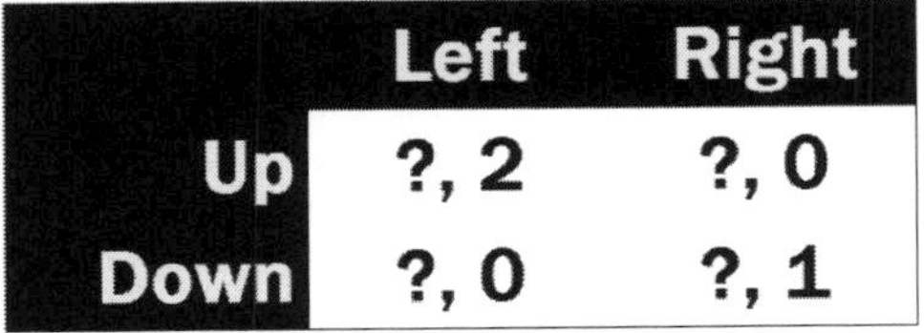

	Left	Right
Up	?, 2	?, 0
Down	?, 0	?, 1

Since the probabilities of each outcome remain the same, it is a simple matter of compiling the probabilities and payoffs:

$EU_1 = (1/3)(2/3)(2) + (2/3)(2/3)(0) + (1/3)(1/3)(0) + (2/3)(1/3)(1)$

$EU_1 = (2/9)(2) + (4/9)(0) + (1/9)(0) + (2/9)(1)$

$EU_1 = 4/9 + 2/9$

$EU_1 = 6/9$

$EU_1 = 2/3$

Why is the mixed strategy Nash equilibrium so bizarre? Both the <ballet, fight> and <fight, ballet> outcomes represent coordination failure. They both occur with positive probability in the MSNE, accounting for 5/9 of the outcomes. That means the couple go on their date less than half of the time if they mix, which drags down their payoffs. Indeed, each would be better off agreeing to meet at their *lesser* preferred form of entertainment; the 1 they earn from that outcome beats the 2/3 they earn in the MSNE. Consequently, the MSNE is a rational but strange set of strategies. In turn, if players ever played the MSNE, we might wonder why they could not simply coordinate on one of the PSNE.

1.6.3: Pure Coordination

Mixed strategies can be even stranger in this regard. Consider the following game:

	Left	Right
Up	1, 1	0, 0
Down	0, 0	1, 1

We call this pure coordination. The players had mixed motives in battle of the sexes; they wanted to be together, but they also wanted to see their preferred form of entertainment. In pure coordination, they only care about being together. One simple interpretation of this is choosing which side of the street to drive on. It does not really matter whether we all drive on the left side or all drive on the right side, as long as some of us do not drive on the left while others drive on the right.

Obviously, <up, left> and <down, right> are PSNE; any deviation from these sets of strategies changes a player's payoff from 1 to 0. However, a MSNE also exists. To solve for it, first consider player 2's payoffs for left as a function of player 1's mixed strategy σ_{up}:

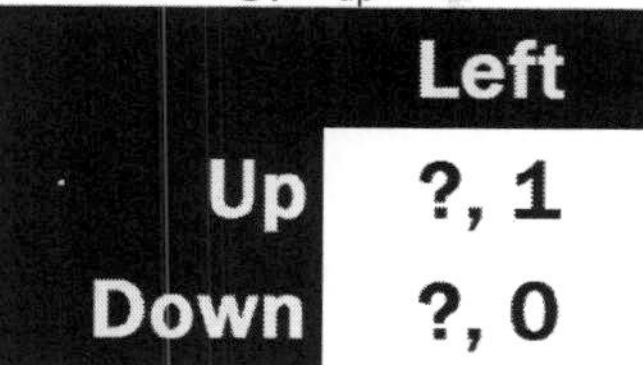

	Left
Up	?, 1
Down	?, 0

She earns 1 with probability σ_{up} and 0 with probability $1 - \sigma_{up}$. In expected utility form:

$$EU_{left} = (\sigma_{up})(1) + (1 - \sigma_{up})(0)$$

Now for right:

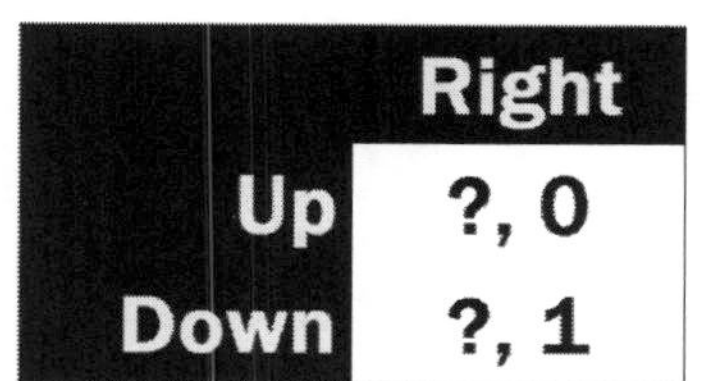

	Right
Up	?, 0
Down	?, 1

This time, she earns 0 with probability σ_{up} and 1 with probability $1 - \sigma_{up}$. Or:

$$EU_{right} = (\sigma_{up})(0) + (1 - \sigma_{up})(1)$$

Putting these together, we solve for σ_{up}:

$$EU_{left} = EU_{right}$$
$$EU_{left} = (\sigma_{up})(1) + (1 - \sigma_{up})(0)$$
$$EU_{right} = (\sigma_{up})(0) + (1 - \sigma_{up})(1)$$
$$(\sigma_{up})(1) + (1 - \sigma_{up})(0) = (\sigma_{up})(0) + (1 - \sigma_{up})(1)$$
$$\sigma_{up} = 1 - \sigma_{up}$$
$$2\sigma_{up} = 1$$
$$\sigma_{up} = 1/2$$

So she mixes equally between left and right. Player 1 does the same with up and down. We can see this by starting with his expected utility for up as a function of σ_{left}:

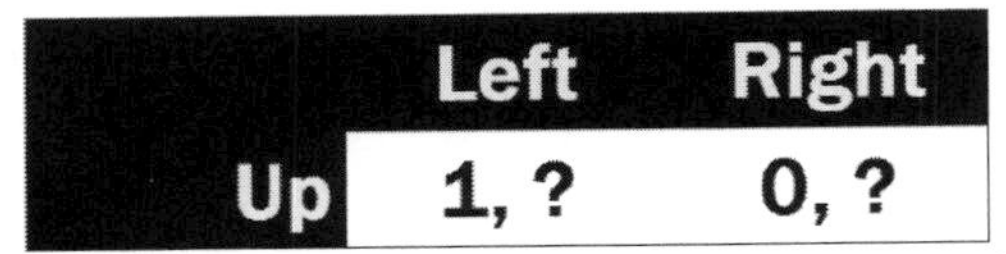

	Left	Right
Up	1, ?	0, ?

Similar to what we saw with player 2, player 1 earns 1 with probability σ_{left} and 0 with probability $1 - \sigma_{left}$. We can write this as:

$$EU_{up} = (\sigma_{left})(1) + (1 - \sigma_{left})(0)$$

And for down:

	Left	Right
Down	0, ?	1, ?

Here, it is 0 with probability σ_{left} and 1 with probability $1 - \sigma_{left}$. Therefore, his expected utility for down equals:

$$EU_{down} = (\sigma_{left})(0) + (1 - \sigma_{left})(1)$$

Now we set EU_{up} equal to EU_{down} and solve for σ_{left}:

$$EU_{up} = EU_{down}$$
$$EU_{up} = (\sigma_{left})(1) + (1 - \sigma_{left})(0)$$
$$EU_{down} = (\sigma_{left})(0) + (1 - \sigma_{left})(1)$$
$$(\sigma_{left})(1) + (1 - \sigma_{left})(0) = (\sigma_{left})(0) + (1 - \sigma_{left})(1)$$
$$\sigma_{left} = 1 - \sigma_{left}$$
$$2\sigma_{left} = 1$$
$$\sigma_{left} = 1/2$$

Since each player selects both of his or her strategies with probability 1/2, each outcome occurs in the MSNE with probability 1/4. In two of these outcomes, the players earn 0. In the other two, they earn 1. As such, their expected utility in the MSNE is 1/2. This is strictly worse than either of the PSNE, which is problematic for the players.

One way to escape the inefficient mixed strategy Nash equilibria in pure coordination and battle of the sexes is to follow social norms and laws. Driving on the road in the United States is very easy because a law tells us to drive on the right side, and that is an efficient Nash equilibrium. In battle of the sexes, perhaps the couple had a rule of thumb that the man chooses where to go on Fridays and the woman chooses where to go on Saturdays. If that were the case, they would only need to look at a calendar to coordinate even if they could not directly communicate. Thus, these strange MSNE help us interpret the usefulness of these types of coordination rules.

1.6.4: A Shortcut for Zero Sum Games

In chicken and battle of the sexes, each player's payoff in the MSNE is equal to his or her opponent's. However, unless the game is as symmetrical as the ones above, that will usually not be the case. To see an example where the payoffs differ, refer back the modified form of matching pennies discussed in Lesson 1.5:

	Left	Right
Up	3, -3	-2, 2
Down	-1, 1	0, 0

Begin by recalling that player 1 mixes with probability 1/6 on up and 5/6 on down, while player 2 mixes with probability 1/3 on left and 2/3 on right. Therefore, <up, left> occurs with probability 1/6 times 1/3; <down, left> occurs with probability 5/6 times 1/3; <up, right> occurs with probability 1/6 times 2/3; and <down, right> occurs with probability 5/6 times 2/3.

Let's focus on player 1's payoffs and use these probabilities to calculate his expected utility:

	Left	Right
Up	3, ?	-2, ?
Down	-1, ?	0, ?

So his expected utility equals:

$EU_1 = (1/6)(1/3)(3) + (5/6)(1/3)(-1) + (1/6)(2/3)(-2) + (5/6)(2/3)(0)$
$EU_1 = (1/18)(3) + (5/18)(-1) + (2/18)(-2) + (10/18)(0)$
$EU_1 = 3/18 - 5/18 - 4/18$
$EU_1 = -6/18$
$EU_1 = -1/3$

Rather than calculating player 2's expected utility as usual, recall that this game is zero sum: every time a player gains some amount, the other player loses that amount. In other words, if we sum the payoffs for each individual outcome, all add up to 0. Given that player 1's gain is player 2's loss and vice versa, if player 1's expected utility equals -1/3, then player 2's expected utility must be the negative of that payoff, or --1/3 = 1/3.

In summary, we have learned two things from this lesson. First, calculating payoffs in mixed strategy Nash equilibria of zero sum games is easy, since you functionally calculate both players' payoffs by finding one player's. And second, the players' payoffs need not be equal in MSNE; "equilibrium" only refers to the stability of certain strategies, not any sort of balance in the players' payoffs.

1.6.5: Checking Your Answer

Recall that the mixed strategy algorithm guarantees that a player earns the same payoff for selecting either of his or her pure strategies. Consequently, we can also calculate a player's payoff by calculating his or her payoff for selecting one of his or her strategies.

For example, we found that player 1 earned -1/3 in the equilibrium of the previous game by going through each possible outcome. But consider his expected utility for choosing up. He receives 3 whenever player 2 goes left and -2 whenever she picks right. In equilibrium, she moves left with probability 1/3 and right with probability 2/3. Therefore, player 1's expected utility for up equals:

$$EU_{up} = (1/3)(3) + (2/3)(-2)$$
$$EU_{up} = 3/3 + -4/3$$
$$EU_{up} = -1/3$$

As claimed, this is his equilibrium expected utility.

Having an additional method to calculate equilibrium expected utilities is useful for a couple reasons. First, it allows you to check your answer. If the methods of calculation produce different payoffs, you know you have done something wrong and need to check things again. Second, although the first method is intuitively easier to grasp, the alternative method is computationally less intensive. Once you feel comfortable with the process, you will want to use this second method exclusively.

Takeaway Points

1) Expected utilities in MSNE are weighted averages of each of the outcomes that occur in equilibrium.
2) In a zero sum game, a player's payoff is the negative of the opposing player's payoff.

Lesson 1.7: Strict Dominance in Mixed Strategies

Consider the following game:

	Left	Right
Up	3, -1	-1, 1
Middle	0, 0	0, 0
Down	-1, 2	2, -1

Specifically, note that middle is not dominated by player 1's other pure strategies. To see this, let's first compare middle to up:

	Left	Right
Up	3, ?	-1, ?
Middle	0, ?	0, ?

If player 2 selects left, player 1 prefers to go up and earn 3, which is better than the 0 he otherwise earns for middle. But if player 2 chooses right, player 1 would now rather go middle, as 0 is greater than the -1 he earns for up. So up does not dominate middle.

Down does not dominate middle either:

	Left	Right
Middle	0, ?	0, ?
Down	-1, ?	2, ?

If player 2 picks left, then player 1 is better off going middle; but if she goes right, he ought to go down. So down does not dominate middle.

Combining these two pieces of information together, we now know that no pure strategy dominates middle. Nevertheless, middle is strictly dominated. How? If a mixture of two pure strategies strictly dominates a third strategy, that third strategy is strictly dominated.

To see how middle is strictly dominated in this particular game, consider a mixed strategy from player 1 where he plays up with probability 1/2 and down with probability 1/2. Let's calculate his expected utility if player 2 plays left:

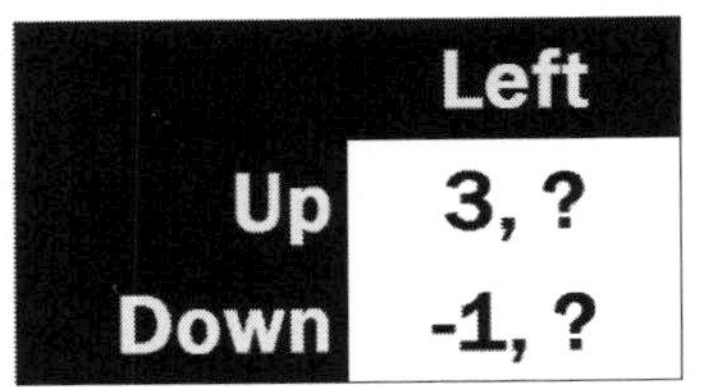

	Left
Up	3, ?
Down	-1, ?

Using this matrix for guidance, we know that he will earn 3 with probability 1/2 and -1 with probability 1/2. As an equation:

(1/2)(3) + (1/2)(-1)
3/2 – 1/2
2/2
1

Note that his expected utility for this mixed strategy is greater than the 0 that he would earn if he played middle as pure strategy.

Now suppose player 2 played right and player 1 stuck to this mixed strategy:

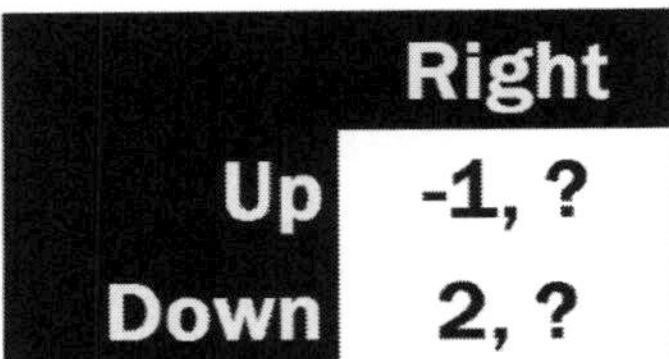

	Right
Up	-1, ?
Down	2, ?

Now he earns -1 with probability 1/2 and 2 with probability 2. Therefore, his expected utility equals:

(1/2)(-1) + (1/2)(2)
-1/2 + 2/2
1/2

Again, this expected utility is greater than the 0 he earns if he plays middle in response to player 2 selecting right. That means regardless of player 2's choice, player 1 would be better off playing this mixture between up and down than playing middle as a pure strategy. Thus, middle is strictly dominated.

Other mixtures between up and down also strictly dominate middle. For example, up with probability 49/100 and down with probability 51/100 or up with probability 51/100 and down with probability 49/100 will both produce a strictly better outcome than middle regardless of what player 2 does. However, this is superfluous information. Once we know that a mixture—any mixture at all—of some pure strategies strictly dominates another pure strategy, we can immediately remove that dominated pure strategy just as we have done in the past.

Specifically, that means we can reduce the original game to this far less intimidating form:

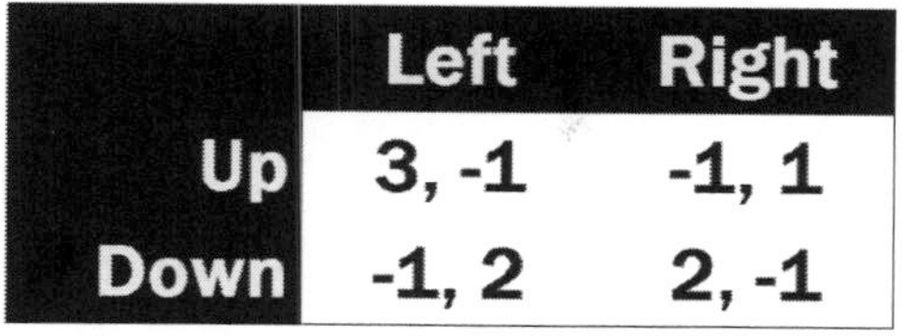

	Left	Right
Up	3, -1	-1, 1
Down	-1, 2	2, -1

It is easy to see that there are no pure strategy Nash equilibria here. For the <up, left> outcome, player 2 can deviate to right and improve from -1 to 1. For the <up, right> outcome, player 1 can deviate to down and improve from -1 to 2. For the <down, right> outcome, player 2 can deviate to left and improve from -1 to 2. Finally, for the <down, left> outcome, player 1 can deviate to up and improve from -1 to 3.

As such, we look to the mixed strategy algorithm to provide a solution. Let's start with player 1's mixed strategy. Consider player 2's payoff for playing left as a pure strategy in response to that mixture:

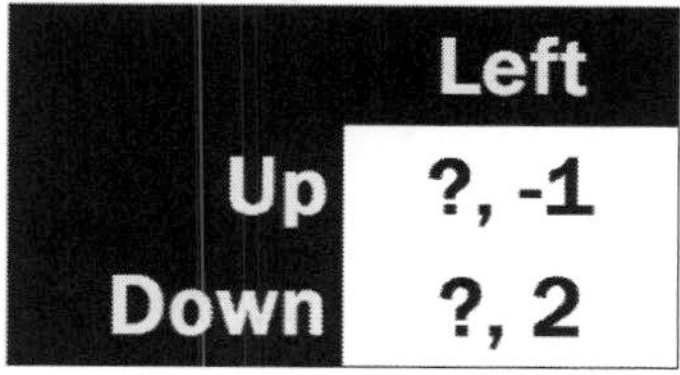

	Left
Up	?, -1
Down	?, 2

So with probability σ_{up}, player 2 earns -1; and with probability $1 - \sigma_{up}$, she earns 2. Therefore, her expected utility equals:

$$EU_{left} = (\sigma_{up})(-1) + (1 - \sigma_{up})(2)$$

Next, we need to find her expected utility for right against that same mixed strategy:

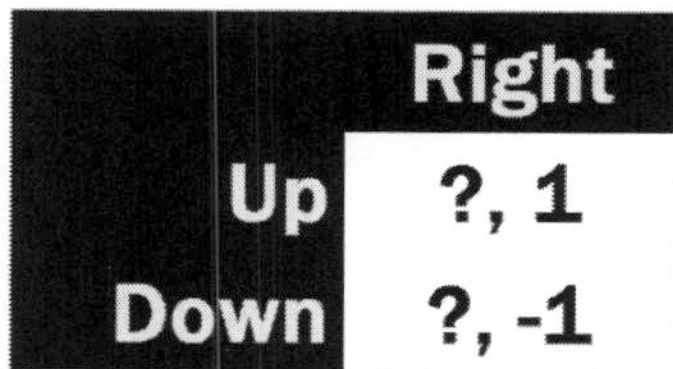

	Right
Up	?, 1
Down	?, -1

Here, she earns 1 with probability σ_{up} and -1 with probability $1 - \sigma_{up}$. As an equation:

$$EU_{right} = (\sigma_{up})(1) + (1 - \sigma_{up})(-1)$$

To finish, we set the expected utility for left equal to the expected utility for right and solve for σ_{up}:

$EU_{left} = EU_{right}$
$EU_{left} = (\sigma_{up})(-1) + (1 - \sigma_{up})(2)$
$EU_{right} = (\sigma_{up})(1) + (1 - \sigma_{up})(-1)$
$(\sigma_{up})(-1) + (1 - \sigma_{up})(2) = (\sigma_{up})(1) + (1 - \sigma_{up})(-1)$
$-\sigma_{up} + 2 - 2\sigma_{up} = \sigma_{up} - 1 + \sigma_{up}$
$-3\sigma_{up} + 2 = 2\sigma_{up} - 1$
$5\sigma_{up} = 3$
$\sigma_{up} = 3/5$

So player 1 picks up with probability 3/5 and down with probability 2/5 in the MSNE.

Let's switch to player 2's mixed strategy. Consider player 1's payoffs for up in response to her mixture:

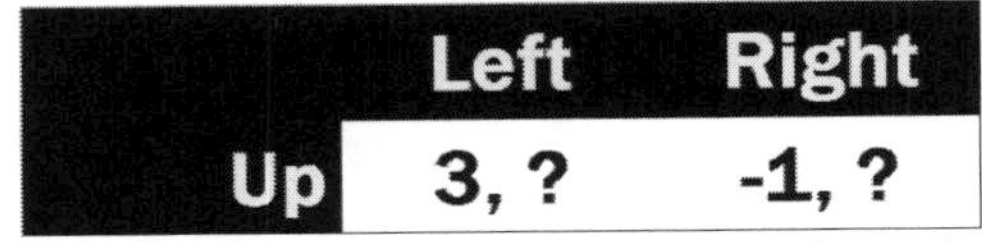

	Left	Right
Up	3, ?	-1, ?

Player 1 earns 3 with probability σ_{left} and -1 with probability $1 - \sigma_{left}$. Or:

$EU_{up} = (\sigma_{left})(3) + (1 - \sigma_{left})(-1)$

Now for down:

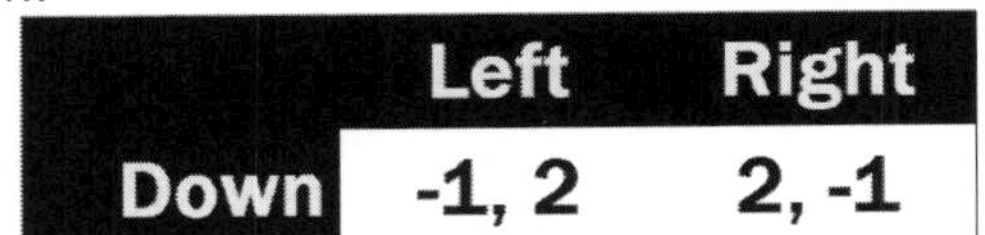

	Left	Right
Down	-1, 2	2, -1

Here, he earns -1 with probability σ_{left} and 2 with probability $1 - \sigma_{left}$. As an equation:

$EU_{down} = (\sigma_{left})(-1) + (1 - \sigma_{left})(2)$

Finally, to solve for the exact mixture, we set his expected utility for up equal to his expected utility for down:

$EU_{up} = EU_{down}$
$EU_{up} = (\sigma_{left})(3) + (1 - \sigma_{left})(-1)$
$EU_{down} = (\sigma_{left})(-1) + (1 - \sigma_{left})(2)$
$(\sigma_{left})(3) + (1 - \sigma_{left})(-1) = (\sigma_{left})(-1) + (1 - \sigma_{left})(2)$

$3\sigma_{left} - 1 + \sigma_{left} = -\sigma_{left} + 2 - 2\sigma_{left}$
$4\sigma_{left} - 1 = -3\sigma_{left} + 2$
$7\sigma_{left} = 3$
$\sigma_{left} = 3/7$

So in the MSNE, player 1 selects up with probability 3/5 and down with probability 2/5 while player 2 chooses left with probability 3/7 and right with probability 4/7.

Although strict dominance guarantees that player 1 cannot profitably deviate to middle, let's verify that this is the case. Specifically, recall that player 1 earns a guaranteed payoff of 0 for selecting middle. Consequently, if player 1 cannot profitably deviate from the MSNE to playing middle as a pure strategy, his expected utility from the MSNE must be at least 0.

To check, let's calculate the probability of each outcome occurring in the MSNE. Given the mixtures up with probability 3/5, down with probability 2/5, left with probability 3/7, and right with probability 4/7, the likelihood that each outcome occurs is simply the respective probabilities multiplied together. That is, <up, left> occurs with probability 3/5 times 3/7; <down, left> occurs with probability 2/5 times 3/7; <up, right> occurs with probability 3/5 times 4/7; and <down, right> occurs with probability 2/5 times 4/7.

Now we match those probabilities with player 1's payoffs for each of the outcomes:

	Left	Right
Up	3, ?	-1, ?
Down	-1, ?	2, -?

We now combine this information together as a formula:

$EU_1 = (3/5)(3/7)(3) + (2/5)(3/7)(-1) + (3/5)(4/7)(-1) + (2/5)(4/7)(2)$
$EU_1 = (9/35)(3) + (6/35)(-1) + (12/35)(-1) + (8/35)(2)$
$EU_1 = 27/35 - 6/35 - 12/35 + 16/35$
$EU_1 = 25/35$
$EU_1 = 5/7$

And, sure enough, his expected utility of 5/7 in the MSNE is better than 0 he would receive for middle as a pure strategy.

1.7.1: Mixed Dominance and IESDS

We can use mixed dominance in a sequence of iterated elimination of strictly dominated strategies as normal. To see this, we will use the following game:

	Left	Center	Right
Up	-3, 6	9, 1	0, 2
Middle	3, -4	2, 4	4, 1
Down	4, 7	3, 2	-3, 2

Here, no pure strategy strictly dominates right for player 2. We can see this by isolating left versus right and center versus right. First, let's compare right with left:

	Left	Right
Up	?, 6	?, 2
Middle	?, -4	?, 1
Down	?, 7	?, 2

Between these two, left is better than right for player 2 if player 1 plays up or down; right is worth 2 for her in each of these cases, while left is worth 6 if he moves up and 7 if he moves down. However, right beats left if player 1 chooses middle, as 1 is greater than -4. As such, left does not strictly dominate right.

Next, compare center to right:

	Center	Right
Up	?, 1	?, 2
Middle	?, 4	?, 1
Down	?, 2	?, 2

Now right is better if player 1 chooses up, by a 2 to 1 margin. In contrast, if player 1 goes middle, player 2's center move earns her 4, which beats the 1 she would earn by choosing right. Finally, if player 1 goes down, she earns 2 regardless of which of these two strategies she chooses. So center does not strictly dominate right.

However, a mixture between left and center does. Let's calculate the expected utility for player 2 if she plays left with probability 1/4 and center with probability 3/4. First, suppose player 1 plays up as a pure strategy:

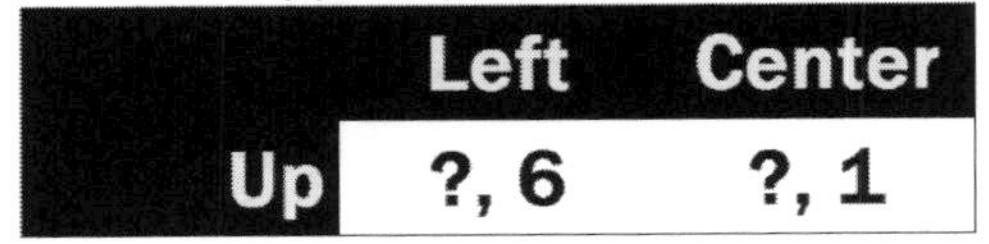

	Left	Center
Up	?, 6	?, 1

So player 2 earns 6 with probability 1/4 and 1 with probability 3/4. Therefore, her expected utility equals:

$EU_2 = (1/4)(6) + (3/4)(1)$
$EU_2 = 6/4 + 3/4$
$EU_2 = 9/4$

In contrast, player 2 earns 2 if she plays right in this situation. Since 9/4 is greater 2, the mixture provides a greater payoff for her than playing right as pure strategy if player 1 selects up.

Now we go to player 1 playing middle as a pure strategy:

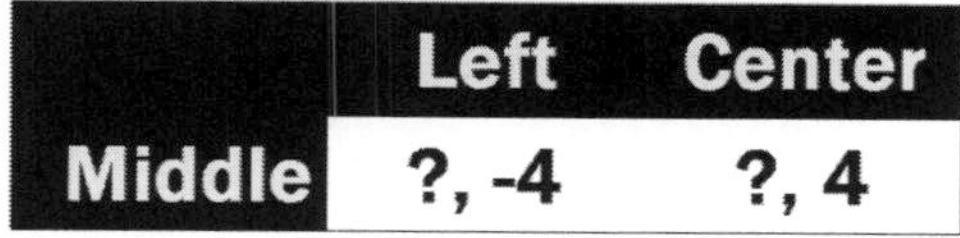

	Left	Center
Middle	?, -4	?, 4

Here, player 2 earns -4 with probability 1/4 and 4 with probability 3/4. As such, her expected utility equals:

$EU_2 = (1/4)(-4) + (3/4)(4)$
$EU_2 = -4/4 + 12/4$
$EU_2 = -1 + 3$
$EU_2 = 2$

Meanwhile, if she plays right in response to player 1 playing middle, she earns 1. Again, the mixture provides a better payoff with 2.

Finally, we move on to player 1 playing down:

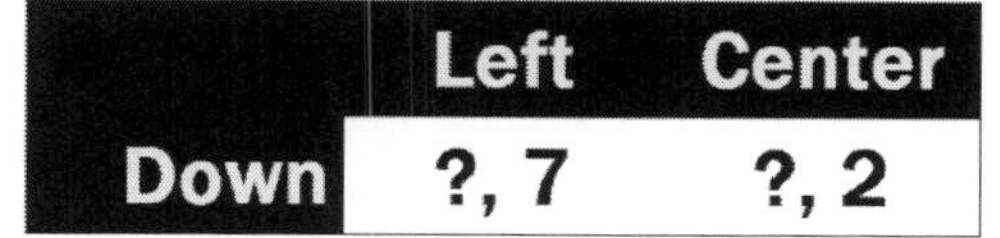

	Left	Center
Down	?, 7	?, 2

In this case, player 2 earns 7 with probability 1/4 and 2 with probability 3/4. Thus, her expected utility equals:

$EU_2 = (1/4)(7) + (3/4)(2)$
$EU_2 = 7/4 + 6/4$
$EU_2 = 13/4$

Playing right as pure strategy earns her 2. Because 13/4 beats 2, player 2 again performs better with the mixed strategy.

Indeed, she always does better with the mixture between left and center than she does with right as a pure strategy, so the mixed strategy strictly dominates right. As such, we can remove it from the matrix, thereby reducing the game to the following:

	Left	Center
Up	-3, 6	9, 1
Middle	3, -4	2, 4
Down	4, 7	3, 2

From here, we can continue our IESDS process as normal. Note that down now strictly dominates middle for player 1:

	Left	Center
Middle	3, ?	2, ?
Down	4, ?	3, ?

That is, 4 beats 3 if player 2 plays left and 3 beats 2 if she plays center. Since down strictly dominates middle, we can remove middle from the matrix and leave the following:

	Left	Center
Up	-3, 6	9, 1
Down	4, 7	3, 2

Left now strictly dominates center for player 2:

	Left	Center
Up	?, 6	?, 1
Down	?, 7	?, 2

If player 1 moves up, left beats center 6 to 1. If he moves down, left wins 7 to 2. Either way, left better serves player 2 than center. Therefore, we can eliminate center, which leaves us with just two outcomes:

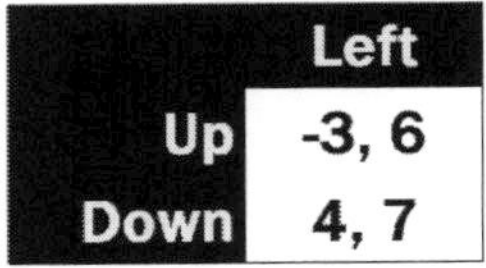

	Left
Up	-3, 6
Down	4, 7

From here, player 1 simply picks the best outcome for himself. Since beats -3, he moves down. Thus, through iterated elimination of strictly dominated strategies, <down, left> is the only Nash equilibrium of this game.

Overall, strict dominance in mixed strategies can be frustrating to work with—there are many combinations of pure strategies and an infinite range of mixtures between those strategies. Consequently, it takes effort to locate such strictly dominant mixed strategies. However, the payoff is ultimately worth it, as we can simplify games a great deal when we do find them.

Takeaway Points

1) Some pure strategies are strictly dominated by mixed strategies but not other pure strategies.
2) We can eliminate such strictly dominated strategies as a normal part of the IESDS process.

Lesson 1.8: The Odd Rule and Infinitely Many Equilibria

Let's think back to some of the games we have solved and how many Nash equilibria they had. The prisoner's dilemma and deadlock each had one pure strategy Nash equilibrium that we could find through strict dominance. Many other games we solved through IESDS also had one. Matching pennies had one in mixed strategies. The stag hunt, chicken, and battle of the sexes each had three: two in pure strategies and one in mixed strategies.

Is it a coincidence that all of these games have an odd number of equilibria? Actually, it is not—a 1971 paper from Robert Wilson showed almost no games have an even or infinite number of equilibria. However, some quirky games do not follow this odd rule of thumb, and our old friend weak dominance frequently claims responsibility. This lesson covers a bunch of such games.

To start, consider the following voting game. Today, I am offering you and your friend $1 each at absolutely not cost to either of you. The two of you only need to approve the transfer to receive the money. Approval requires a unanimous positive vote on a secret ballot. As such, the two of you will simultaneously cast your votes. I will read the ballots and distribute the money if the resolution passes.

At first thought, we might think that the only reasonable outcome is for each of you to receive $1. However, voting games like this can have unexpected outcomes. After all, the resolution requires unanimity. As such, if both of you reject the offer, neither of you can individually switch to accepting and alter the outcome. This leads to an inefficient Nash equilibrium.

To understand the logic here, let's look at the game's matrix. Imagine up and left represent accept while down and right represent reject:

	Left	Center
Up	1, 1	0, 0
Down	0, 0	0, 0

Note that up weakly dominates down:

	Left	Center
Up	1, ?	0, ?
Down	0, ?	0, ?

That is, if player 2 plays left, up beats down; but if player 2 plays right, up and down are equally good.

Likewise, left weakly dominates right:

	Left	Center
Up	?, 1	?, 0
Down	?, 0	?, 0

Here, if player 1 plays up, left is better than right; but if player 1 plays down, left and right are equally good.

Through iterated elimination of weakly dominated strategies, <up, left> is a pure strategy Nash equilibrium. This should not come as a surprise given that <up, left> provides the greatest payoff for both players.

However, <down, right> is also a PSNE. To see this, suppose player 1 played down. Then no matter what player 2 does, she earns 0. As such, right is a best response to down. Similarly, if player 2 is playing right, player 1 earns 0 whether he plays up or down. Therefore, down is a best response to right. That means if the players choose <down, right> they cannot individually improve their payoffs to a value greater than zero. Thus, <down, right> is a mutual best response and a Nash equilibrium, which means it is reasonable for the two of you to walk away without any money.

So far, we have found two Nash equilibria. Are there any others? Clearly <down, left> and <up, right> are not. In the case of <down, left>, player 1 can profitably deviate to up and increase his payoff from 0 to 1; likewise, in the case of <up, right>, player 2 can profitably deviate to left and increase her payoff from 0 to 1. So if this game only has two Nash equilibria, there must be no mixed strategy Nash equilibrium.

To verify the absence of a MSNE, suppose player 1 mixed with probability σ_{up}. Then player 2's payoff for playing left is:

$$EU_{left} = (\sigma_{up})(1) + (1 - \sigma_{up})(0)$$
$$EU_{left} = \sigma_{up}$$

And her payoff for playing right is:

$$EU_{right} = (\sigma_{up})(0) + (1 - \sigma_{up})(0)$$
$$EU_{right} = 0$$

Consequently, if $\sigma_{up} > 0$ (that is, as long as player 1 is truly mixing and not playing down with certainty), player 2's best response to any such mixture is to play left as pure strategy. The reason is simple—if she plays left, there is some chance she will receive 1, but she is guaranteed to earn 0 if she plays down.

Player 1's best response to a mixture from player 2 works the same way. Suppose player 2 mixed with probability σ_{left}. Then player 1's payoff for selecting up equals:

$EU_{up} = (\sigma_{left})(1) + (1 - \sigma_{left})(0)$
$EU_{up} = \sigma_{left}$

And his payoff for playing down is:

$EU_{down} = (\sigma_{left})(0) + (1 - \sigma_{left})(0)$
$EU_{down} = 0$

So just as before, if player 2 puts positive probability on left and right, then player 1's best response is to always play up. As a result, the players cannot mix, thus making this a rare game that has an even number of equilibria.

1.8.1: Infinitely Many Equilibria

Games can also have infinitely many equilibria. We will look at two different ways this can occur. First, consider this game:

	Left	Center
Up	2, 2	9, 0
Down	2, 3	5, -1

To solve it, begin by noting that left strictly dominates right:

	Left	Center
Up	?, 2	?, 0
Down	?, 3	?, -1

If player 1 selects up, left beats right for player 2 by a 2 to 0 margin. If he plays down, left defeats right 3 to -1. So regardless of what player 1 does, player 2 ought to move left. Thus, through iterated elimination of strictly dominated strategies, we can reduce the game to this:

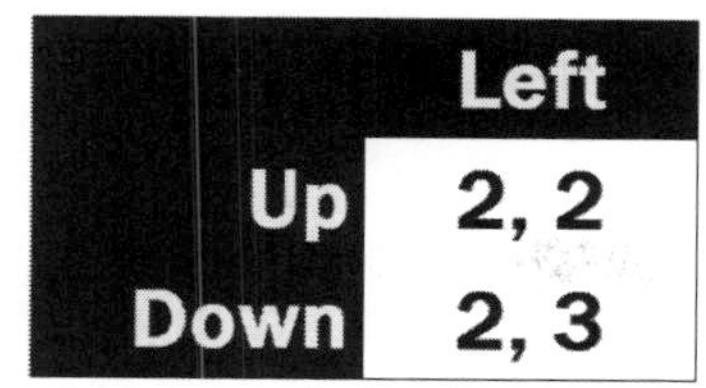

	Left
Up	2, 2
Down	2, 3

Knowing that player 2 chooses left, player 1 earns 2 no matter what he does. Therefore, both up and down are pure strategy best responses to left. In turn, <up, left> and <down, left> are pure strategy Nash equilibria.

While that should be obvious, a subtle implication here is that *any* mixture between up and down is also a best response to left. To see this, consider player 1's expected utility for any mixed strategy. We know player 2 selects left, so player 1's payoff is as follows:

$EU_1 = (\sigma_{up})(2) + (1 - \sigma_{up})(2)$

That is, he earns 2 with probability σ_{up} (whatever percentage of the time he chooses to play up) and 2 with probability $1 - \sigma_{up}$ (whatever percentage of the time he chooses to play down). However, this formula quickly simplifies to 2:

$EU_1 = (\sigma_{up})(2) + (1 - \sigma_{up})(2)$
$EU_1 = 2\sigma_{up} + 2 - 2\sigma_{up}$
$EU_1 = 2$

Since this is equal to his payoff for playing up or down pure strategy, all of these mixtures are also best responses to left. As such, in equilibrium, player 1 can play up as pure strategy, down as a pure strategy, or mix in any fashion between those two. Infinitely many of such mixtures exist; therefore, this game has infinitely many equilibria. Although this falls under the MSNE umbrella, we can be very specific and call this a *partially* mixed strategy Nash equilibrium, as player 1 mixes but player 2 does not.

Another interesting thing to note is that player 1 plays a weakly dominated strategy with positive probability in the MSNE. To see this, let's focus on player 1's payoffs in isolation:

	Left	Center
Up	2, ?	9, ?
Down	2, ?	5, ?

As we have just seen, if player 2 selects left, both up and down are equally as good. But putting right back into the matrix reveals that up is better for player 1 in that case. Consequently, up weakly dominates down. However, if we eliminated down, the <down, left> PSNE and the set of MSNE where player 1 mixes between up and down with any probability both disappear. In case the point about weak dominance was not already clear, this game provides another example of why we have to be extremely careful about eliminating weakly dominated strategies.

Moving on, a game can have infinitely many equilibria in a different way. The previous example was trivial because player 1's best response to player 2's strategy was uninteresting—he earned 2 no matter what he did. No change in his strategy would make player 2 want to change her strategy. But that is not the case in all games. Recall back to this game, which we saw in Lesson 1.4:

	Left	Right
Up	2, 3	4, 3
Down	3, 3	1, 1

After marking best responses, we saw that <down, left> and <up, right> were both pure strategy Nash equilibria:

	Left	Right
Up	2, 3*	4*, 3*
Down	3*, 3*	1, 1

Note that both left and right are best responses to up for player 2. Consequently, as long as player 1 selects up as a pure strategy, player 2 is free to mix between left and right in any combination. Given that, we should wonder what sorts of mixed strategies employed by player 2 would induce player 1 to play up over down.

Unlike the previous game, this requires a few calculations. First, we need to know player 1's payoff for playing up as a function of player 2's mixed strategy:

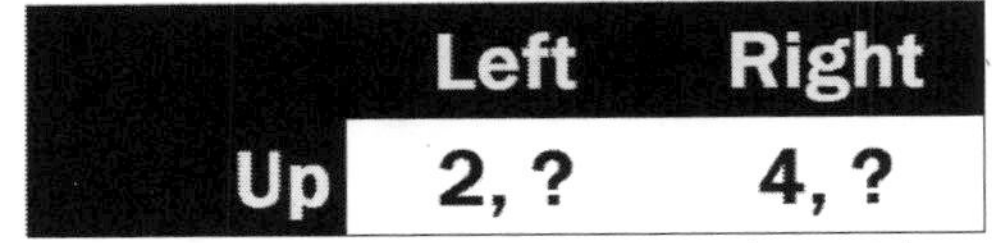

	Left	Right
Up	2, ?	4, ?

So player 1 earns 2 with probability σ_{left} and 4 with probability $1 - \sigma_{left}$. We can write that as this equation:

$EU_{up} = (2)(\sigma_{left}) + (4)(1 - \sigma_{left})$

Moving on to down:

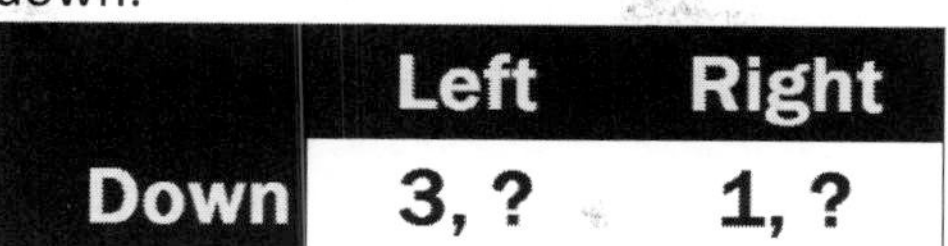

	Left	Right
Down	3, ?	1, ?

Now player 1 earns 3 with probability σ_{left} and 1 with probability $1 - \sigma_{left}$. As an equation:

$EU_{down} = (3)(\sigma_{left}) + (1)(1 - \sigma_{left})$

Remember that a strategy is a best response if it provides at least as good of a payoff as any other alternative. Thus, playing up is a best response to player 2's mixed strategy if the expected utility of playing up is greater than or equal to player 1's expected utility for playing down. Bringing together the previous two equations and solving for the inequality yields the following:

$EU_{up} \geq EU_{down}$
$EU_{up} = (2)(\sigma_{left}) + (4)(1 - \sigma_{left})$
$EU_{down} = (3)(\sigma_{left}) + (1)(1 - \sigma_{left})$
$(2)(\sigma_{left}) + (4)(1 - \sigma_{left}) \geq (3)(\sigma_{left}) + (1)(1 - \sigma_{left})$
$2\sigma_{left} + 4 - 4\sigma_{left} \geq 3\sigma_{left} + 1 - \sigma_{left}$
$-2\sigma_{left} + 4 \geq 2\sigma_{left} + 1$
$4\sigma_{left} \leq 3$
$\sigma_{left} \leq 3/4$

From this, we know that up is a best response for player 1 if player 2 plays left no greater than 3/4 of the time (and thus plays right at least 1/4 of the time). Essentially, such a mixture successfully deters player 1 from choosing down, thereby securing player 2 her payoff of 3. And since player 2 earns 3 regardless of what she does if player 1 plays up, she too is playing a best response to his strategy. Therefore, infinitely many Nash equilibria exist: <down, left>, <up, right>, and infinitely many in mixed strategies where player 1 selects up and player 2 chooses left with probability no greater than 3/4.

Once more, weak dominance is responsible; left weakly dominates right for player 1:

	Left	Right
Up	?, 3	?, 3
Down	?, 3	?, 1

We already know player 2 is indifferent between left and right if player 1 plays up, but left beats right for her if he moves down. Absent that weak dominance, the infinitely many equilibria disappear.

Here is a similar game:

	Left	Right
Up	3, 1	0, 0
Down	2, 2	2, 2

There are two pure strategy Nash equilibria: <up, left> and <down, right>. We can verify this by marking best responses. To begin, suppose player 1 went up:

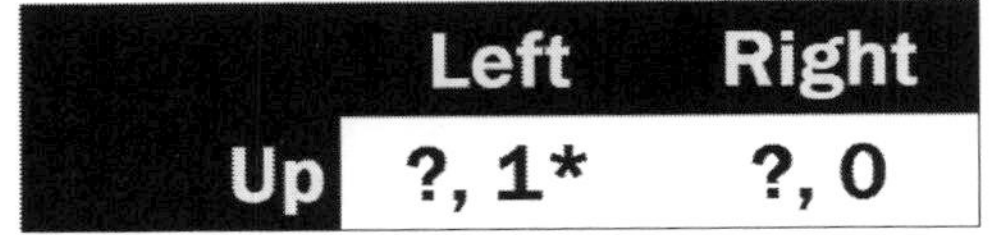

	Left	Right
Up	?, 1*	?, 0

Since 1 beats 0, left is player 2's best response to up.

On the other hand, suppose player 1 went down:

	Left	Right
Down	?, 2*	?, 2*

Now player 2 earns 2 regardless of what she does. As such, both left and right are optimal.

Let's move on to player 1's best responses. To start, suppose player 2 played left:

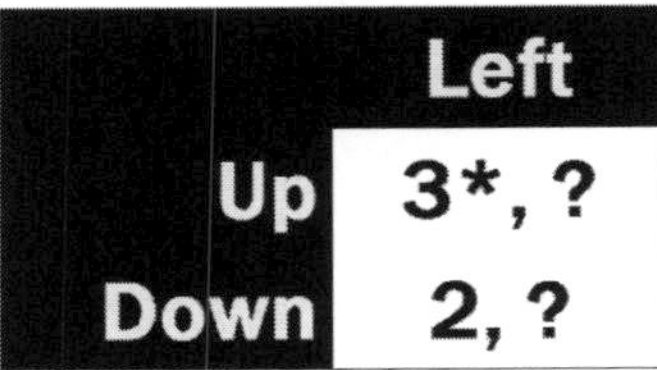

	Left
Up	3*, ?
Down	2, ?

Then up is the best response, as 3 beats 2.

Finally, consider what happens if player 2 plays right:

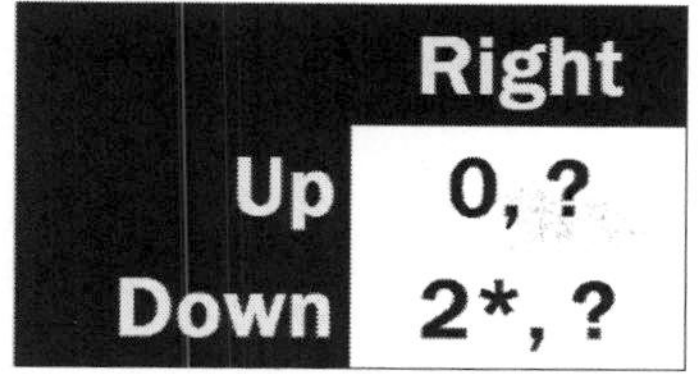

	Right
Up	0, ?
Down	2*, ?

Since 2 beats 0, down is the best response here.

Putting all of that together, the original game looks like this:

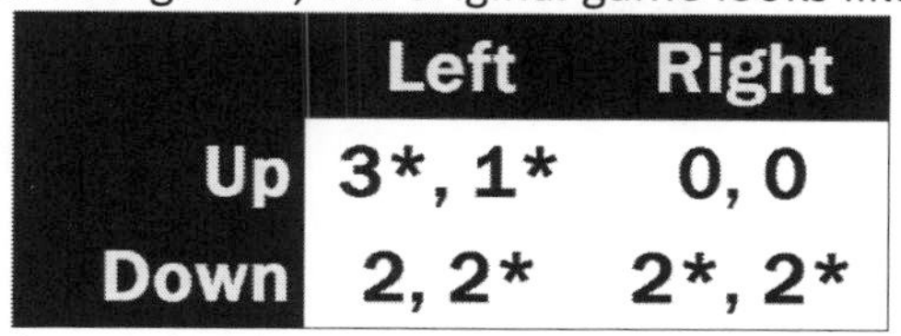

	Left	Right
Up	3*, 1*	0, 0
Down	2, 2*	2*, 2*

Since <up, left> and <down, right> have asterisks for both players payoffs, they are mutual best responses and therefore pure strategy Nash equilibrium.

However, notice that left and right are both best responses for player 2 if player 1 goes down. As such, if player 1 selects down as a pure strategy, player 2 can mix freely between left and right. Consequently, we need to check if down is a best response for player 1 to any of these mixtures from player 2.

To do so, we must write out player 1's expected utility for each of his pure strategies as a function of player 2's mixed strategy. Let's begin with up:

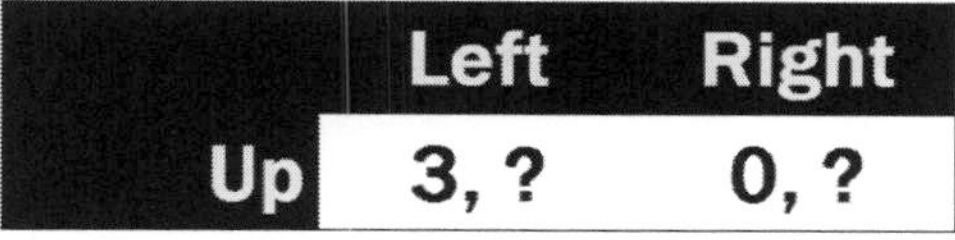

	Left	Right
Up	3, ?	0, ?

In this case, player 1 earns 3 with probability σ_{left} and 0 with probability $1 - \sigma_{left}$. As an equation:

$EU_{up} = (3)(\sigma_{left}) + (0)(1 - \sigma_{left})$
$EU_{up} = 3\sigma_{left}$

Next, we must check down:

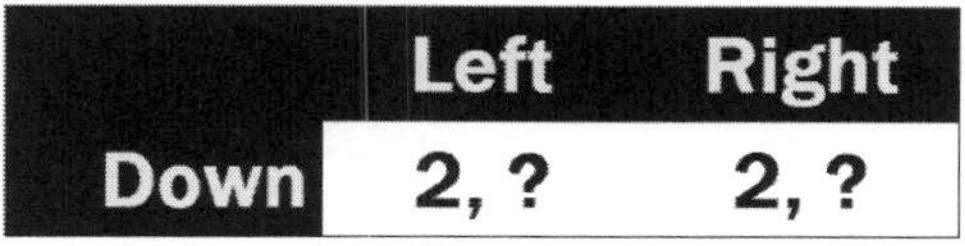

	Left	Right
Down	2, ?	2, ?

Here, he earns 2 with probability σ_{left} and 2 with probability $1 - \sigma_{left}$. Unsurprisingly, that simplifies to just 2. So regardless of player 2's mixed strategy, player 1 earns 2 if he selects down.

Recall that we are looking for all mixed strategies for player 2 that make down a best response for player 1. That is, we need to know for what values of σ_{left} is player 1's expected utility for down greater than or equal to his expected utility for up. A little bit of algebra yields the answer:

$EU_{up} \leq EU_{down}$
$EU_{up} = 3\sigma_{left}$
$EU_{down} = 2$
$3\sigma_{left} \leq 2$
$\sigma_{left} \leq 2/3$

Ergo, this game has infinitely many mixed strategy Nash equilibria in which player 1 plays down as a pure strategy and player 2 chooses left with probability no greater than 2/3 and plays right with the remaining probability.

Once more, weak dominance is at work:

	Left	Right
Up	?, 1	?, 0
Down	?, 2	?, 2

Left weakly dominates right. If player 1 moves down, then player 2 earns 2 regardless of her move. But if player 1 chooses up, the 1 for left trumps the 0 for right.

1.8.2: Take or Share?

Let's look at an application that involves infinitely many equilibria. There is a pot of $8,000. Simultaneously, both players select whether to take or share the money. If both share, they split the pot evenly. If one takes and the other attempts to share, the taker steals all of the money. Finally, if both take, no one receives any money.

Suppose the players only want to maximize their share of the money. Then here is a matrix representation of the game:

	Share	Take
Share	4, 4	0, 8
Take	8, 0	0, 0

This game has appeared on various reality shows with increasing regularity and has been a staple of game shows. (If you have ever seen

Friend or Foe, *Golden Balls*, *Shafted*, or *The Bank Job*, each show's bonus round uses these rules.)

Many people identify this as a prisoner's dilemma and claim <take, take> is the unique equilibrium of this game. However, this game is far more complicated and has many more Nash equilibria.

Let's begin by finding each player's best responses. To start, consider player 1's response to player 2 sharing:

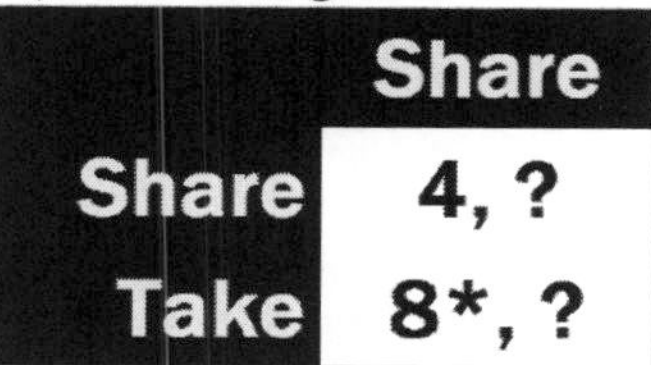

	Share
Share	4, ?
Take	8*, ?

If player 1 takes, he steals all $8,000. If he shares, he only receives half that. Thus, take is the best response.

Now suppose player 2 takes:

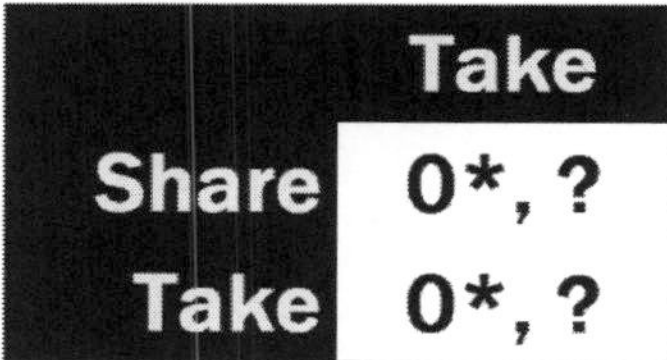

	Take
Share	0*, ?
Take	0*, ?

Here, he earns $0 regardless of his selection. As such, each 0 receives an asterisk.

The same is true for player 2. First, suppose player 1 shares:

	Share	Take
Share	?, 4	?, 8*

Just as before, player 2 doubles her winnings by taking, so the 8 earns the star.

Now consider player 1 taking:

	Share	Take
Take	?, 0*	?, 0*

This time, player 2 is stuck with 0, so both receive the asterisk.

Let's put all of that together:

	Share	Take
Share	4, 4	0*, 8*
Take	8*, 0*	0*, 0*

Already, we see how vastly different take or share is from the prisoner's dilemma. The prisoner's dilemma had a unique equilibrium, as confess was a strictly dominant strategy for both players. Here, take *weakly* dominates share, which allows for <take, share> and <share, take> to also be Nash equilibria.

Delving further, we can show there are infinitely many Nash equilibria. The key is that a player is indifferent between taking and sharing if the other player takes. Thus, the player not taking is free to randomize between taking and sharing.

The only question remaining is whether the taking player is still willing to take as a pure strategy against that randomization. Specifically, it must be that the taking player's expected utility for taking is greater than or equal to his or her expected utility for sharing. But take weakly dominates share, so it must be the case that his or her expected utility for taking is greater than or equal to sharing. So taking as a pure strategy is a best response.

That leaves us with a host of equilibria. First, there are three in pure strategies: <take, take>, <share, take>, and <take, share>. Then there are infinitely many in partially mixed strategies. In these partially mixed strategy Nash equilibria, exactly one player takes as a pure strategy while the other mixes freely between take and share.

As always, keep in mind that these equilibria are a function of the preferences of the players. In this case, we assumed that the players only wanted to maximize money. However, they may have other motivations. For example, if they are slightly vengeful, they may want to maximize their share of the money *and* want to deny the other player any money if the opponent chooses take. Such preferences make taking a strictly dominant strategy, which turns this game into a regular prisoner's dilemma. Meanwhile, generous players may want to mimic the other player's strategy; that is, they want to share if their opponent wishes to share and take if their opponent wishes to take. These preferences mirror the stag hunt, which has three equilibria and allows for cooperative play.

In any case, whenever you look at a model, you should always question the players' preferences. Are players really indifferent between sharing and taking when their opponents take in the take or share game? Do players really always want to maximize money? Does anyone have benevolent preferences?

Game theoretical models can use all of these preferences, but it is up to the game theorist to build them into the structure of the game. If you believe a model has bizarre preferences, consider altering the payoffs and solve the new game for yourself. Your answers will be different and potentially better match reality.

Takeaway Points

1) Almost all games have an odd number of equilibria.
2) Weak dominance often leads to violations of the rule.
3) If you find an even number of equilibria, double check your work to make sure you are not missing any.

Lesson 2.1: Game Trees and Subgame Perfect Equilibrium

At the end of the last chapter, we looked at this game:

	Left	Right
Up	3, 1	0, 0
Down	2, 2	2, 2

We call this Selten's game, named after Reinhardt Selten, a Nobel Prize winning game theorist. Selten contributed to the theory of equilibrium selection. He claimed that certain equilibria make more sense than others given the context of the game.

Recall that Selten's game has infinitely many equilibria. Two are in pure strategies: <up, left> and <down, right>. The rest are a continuum of partially mixed strategy Nash equilibria, in which player 1 selects down as a pure strategy and player 2 chooses left with probability no greater than 2/3 and right with remaining probability.

In terms of prediction, multiple equilibria are problematic. It would be nice if we could bluntly predict that player 1 always takes some action and player 2 always takes another action. Here, however, we must remain vague. If the matrix form of Selten's game modeled a real world phenomenon, player 1 *could* play up and player 2 *could* play left in equilibrium. Or player 1 *could* play down and player 2 *could* play right. Or they *could* play the partially mixed strategy Nash equilibrium. All told, we cannot say which of these the players actually choose purely based off math.

That said, matrix games assume players move at the same time or cannot see each other's moves. We know of many examples where such an assumption makes sense. When prisoners are sequestered in interrogation rooms, they cannot observe what the other one has done. In American football, the offense does not know what type of play the defense has called when it makes its play, and the defense does not know what type of play the offense has called when it makes its play. When a couple tries to meet at the same location without the ability to communicate, they effectively select their destinations with the other side in the dark.

However, some strategic interactions flow over time in specific steps. We call these types of games *sequential* games, since the order of play follows a sequence. For example, a country's army might invade an island, and then its rival must decide whether to attack or concede the territory. A police officer may request to search a suspect's vehicle but must wait for permission before deciding how thoroughly to search. In the game of nim, players take turns selecting how many objects to remove from a pile; the player to pick up the last object wins. In chess, white takes a turn, black replies, and the cycle repeats.

We have seen Selten's game as a simultaneous move game. But what if the players moved sequentially? Consider the following scenario. Firm 2 currently holds a monopoly on a particular market. Firm 1 is considering whether to challenge Firm 2's monopoly status. If it enters, Firm 2 must decide whether to accede to Firm 1's entry or declare a price war. If Firm 2 declares a price war, all the profits go away, and both earn 0. If Firm 2 accedes, both firms can profit. Here, Firm 1 receives a payoff of 3 while Firm 2 receives a payoff of 1. If Firm 1 stays out, it saves its investment and receives a payoff of 2. Meanwhile, without the competition of Firm 1, Firm 2 can increase its payoff to 2.

We normally express such interactions using game trees:

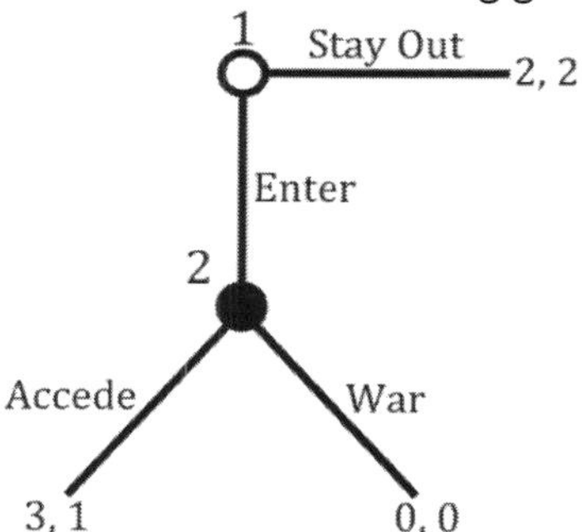

We call this the *extensive form* of the firm entry game. The interaction begins at the open circle—called a decision node—where Firm 1 chooses whether to enter or stay out. Firm 2 selects accede or war at her decision node only if Firm 1 enters.

Notice that the moves match the payoff matrix of Selten's game. Specifically, Firm 1's stay out move is equivalent to down, as players earn 2. Meanwhile, Firm 1's enter move is equivalent to up. It then allows Firm 2 to choose between accede (left) and war (right). The payoffs match the <up, left> and <up, right> outcomes. As such, we could rewrite Selten's game as follows:

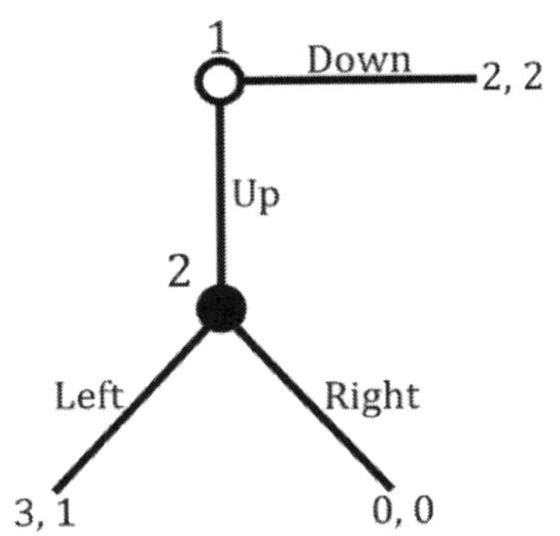

Do the pure strategy Nash equilibria from Selten's game still make sense if we think of the game sequentially, as in the firm entry example? The <up, left> equilibrium certainly does. If player 2 ever had a chance to move, she would pick left over right. Looking at her decision node makes this obvious:

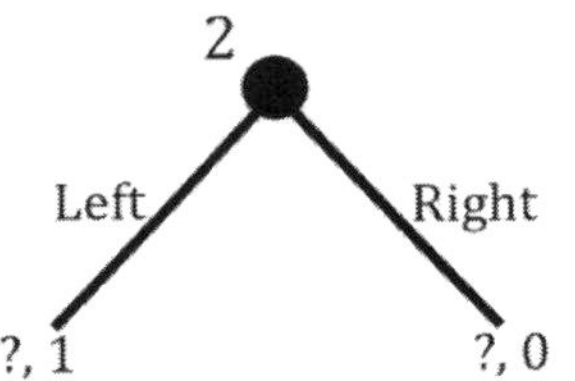

We call this portion of the game a *subgame*. Since player 2 can observe player 1's move, she knows he has selected up. At this point, she only needs to worry about maximizing her payoff. If she moves left, she earns 1; if she chooses right, she earns 0. Therefore, she chooses left. Meanwhile, player 1 receives a payoff of 3, which is his largest expected utility possible. Neither has any incentive to deviate from their strategies, so the <up, left> Nash equilibrium remains sensible.

What about the <down, right> equilibrium? Suppose player 2 committed herself to playing right. Consider player 1's dilemma:

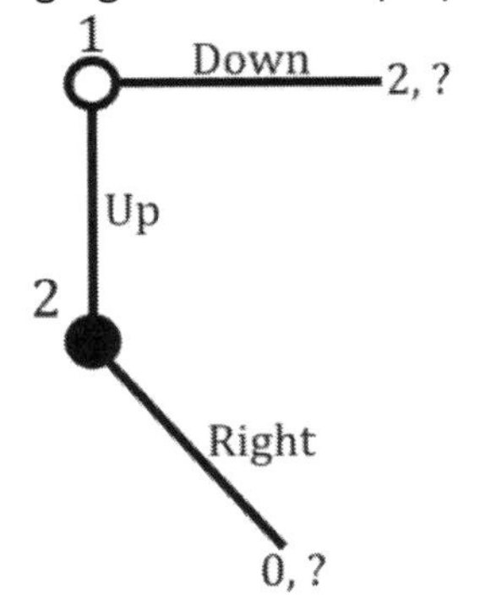

If player 1 chooses down, he earns 2. If he selects up, player 2 moves right, and he earns 0. Since 2 is greater than 0, player 1 would want to select down.

Given that, player 2 would not want to change her strategy. If player 1 moves down, player 2 earns 2, which is her best possible payoff. So <down, right> might seem like a reasonable equilibrium as well.

But is it? Suppose player 1 ignored player 2's threat to move right and selected up anyway. Player 2 must now choose between left and right:

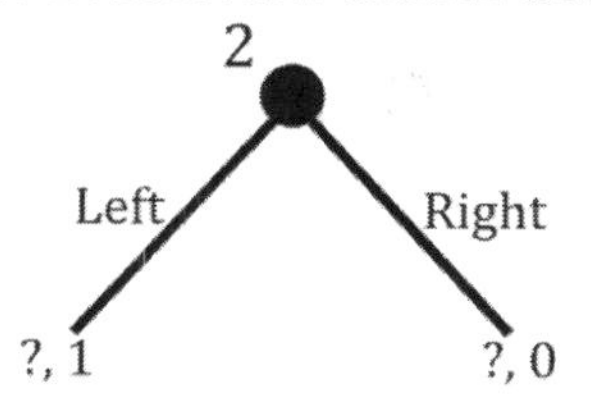

Although player 2 threatened to play right, as soon as player 1 moves up, player 2 has no incentive to follow through on that threat. If she does, she earns 0; if she plays left, she earns 1. As such, player 2 has a profitable deviation to play left. Player 1 recognizes player 2's vulnerability and therefore selects up. In turn, player 2 selects left. As such, we cannot reasonably believe the players will play the <down, right> equilibrium.

Note that this implies that player 2 cannot reasonably play the partially mixed strategy Nash equilibrium either. When we looked at the simultaneous move game, we found an equilibrium in which player 2 played left with probability no greater than 2/3. However, if player 1 moves up, the only rational thing player 2 can do is select left. This precludes the possibility of mixing between left and right, which destroys the partially mixed strategy Nash equilibrium as well.

Thus, only the <up, left> Nash equilibrium survived when we looked at the game sequentially. We call such an equilibrium a *subgame perfect equilibrium* (SPE). Subgame perfection ensures that players only believe threats that others have incentive to carry out when it is time to execute those threats. Here, player 2 threatened to play right, but she could not follow through once it was her turn to move. Thus, player 1 refused to believe her threat to move right, leaving <up, left> as the only Nash equilibrium that survived subgame perfection.

Just as Nash equilibrium is the gold standard for simultaneous move games, subgame perfect equilibrium is the gold standard for extensive form games. As we saw in this example, all SPE are Nash equilibria, but not all Nash equilibria are SPE. As such, subgame perfection is a refinement of Nash equilibrium to ensure that players' threats are credible.

Consequently, we will be working with SPE for the remainder of this chapter.

2.1.1: The Meaning of the Numbers

In the first chapter, we discussed how the payoffs represented a player's subjective ranked ordering of possible outcomes, with the largest number representing the best outcome and the smallest outcome representing the worst. While all that remains true in extensive form games, we now assume that the payoffs represent a ranked ordering of outcomes given what has happened in the game. Consequently, the payoffs reflect a player's evaluation of fairness, distributive justice, and equality.

Some critics of game theory claim that Selten's game makes a false prediction. They argue that, in real life, someone in player 2's shoes would be bitter that player 1 was not "nice" and did not select down. In turn, to punish player 1, player 2 would select right and force player 1's payoff to 0, even though this reduces player 2's payoff from 1 to 0.

However, this criticism is ignorant of our game theoretical principles. Let's isolate player 2's decision once again:

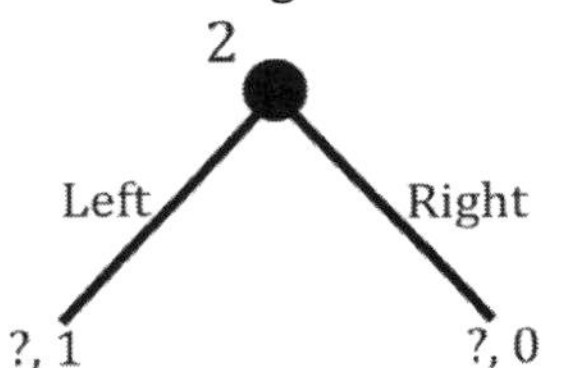

As mentioned, payoffs incorporate players' subjective views on fairness, distributive justice, and equality by definition. Thus, these expected utilities tell us that if player 1 moves up, player 2 would rather move left than right. To say that player 2 would still rather move right to punish player 1 double counts her view of fairness. If she actually valued fairness in the manner described, her payoff for left would be 0 and her payoff for right would be 1. But that is not the case—she earns 1 for left and 0 for right. Thus, if player 1 moves left, we *know* player 2 moves right.

As we discussed in the previous chapter, game theory does not normatively tell players how to think or what their preferences should be. Indeed, we can model scenarios where players value fairness more than their own financial well being. But, when we do, the payoffs *already incorporate these types of preferences*. Do not over think the game; accept the numbers as they appear.

2.1.2: Games with Simultaneous Moves

In Selten's game, the players took turns moving. However, some extensive form games involve simultaneous moves. Here is a simple example:

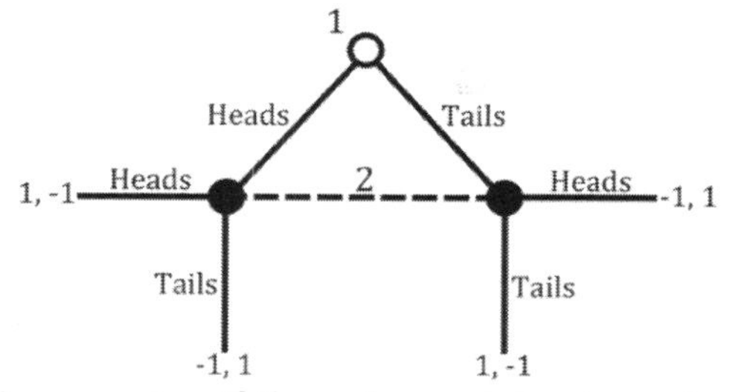

This is matching pennies. If the coins match, player 1 earns 1 and player 2 earns -1. Otherwise, player 1 earns -1 and player 2 earns 1. Player 1 begins by choosing heads or tails. Player 2, then chooses heads or tails without seeing player 1's move. The dashed line indicates that player 2 is blind to player 1's strategy. We call this dashed line player 2's *information set*. The information it conveys is that player 1 played heads or tails, but she cannot see which.

We cannot use the method we saw in Selten's game to solve this game. Previously, we deciphered player 2's optimal move and based player 1's initial decision off of that knowledge. However, player 2 does not know what player 1 did when she moves. She is in the dark, just like in the simultaneous move version of the game.

In fact, when we encounter simultaneous moves in extensive form games, the best thing to do is convert that game to a matrix and solve the game. Notice that the following matrix game has the same strategies and outcomes as the extensive form did:

	Left	Right
Up	1, -1	-1, 1
Down	-1, 1	1, -1

We already know that both players to mix between heads and tails each with probability .5. Thus, no additional work is necessary here; those mixed strategies are also the equilibrium of the extensive form version of the game.

2.1.3: Constructing Games with Simultaneous Moves

Before moving on, there are a couple of important points about simultaneous moves in extensive form games. First, notice that player 2's strategies are identical regardless of how player 1 began the game:

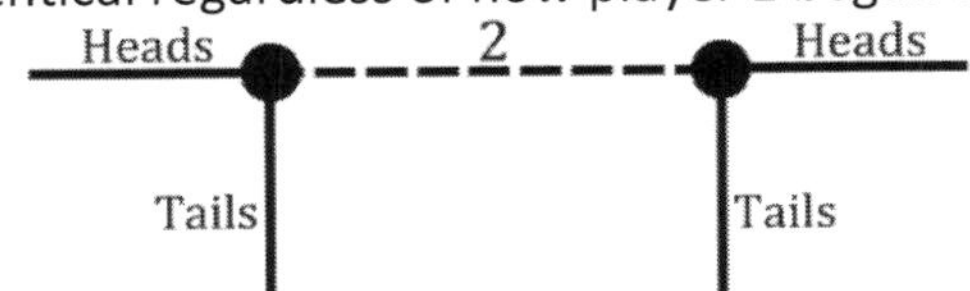

This must be the case whenever there is a simultaneous move and we use the dashed line. To see why, consider the alternative:

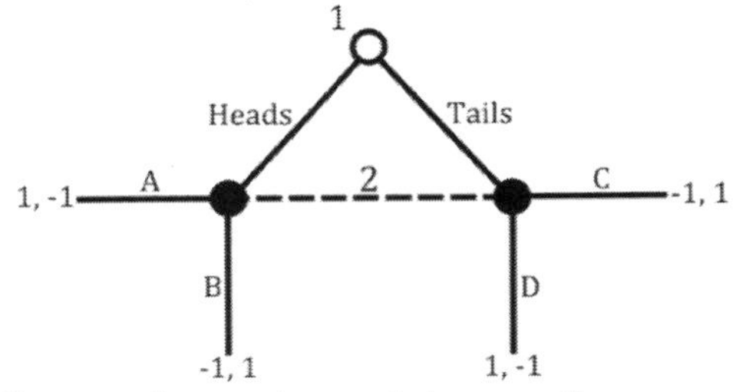

Here, depending on how player 1 began the game, player 2 selects from different strategies. If he chose heads, player 2 chooses between A and B; if he chose tails, she chooses between C and D. However, this has an odd implication. Consider player 2's thought process. She is not supposed to know whether player 1 chose heads or tails. But after player 1 moves, she sees the following:

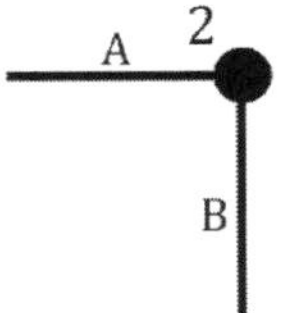

She can now choose between A or B. But this gives away player 1's initial move! She knows the only way she could play A or B is if player 1 selects heads. So if she sees that A and B are her choices, she can infer player 1's original strategy was heads. That ruins the simultaneous nature of the game, which is why game trees must have identical strategies after simultaneous moves.

Moreover, player 2 cannot have an extra strategy depending on player 1's move. To see why, consider this slightly modified version of matching pennies:

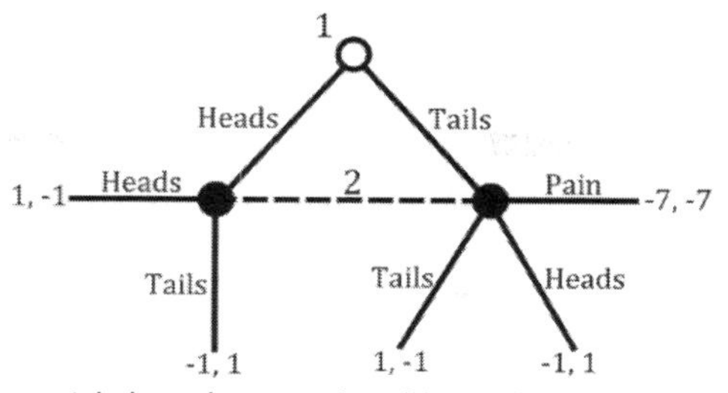

Player 1 begins with heads or tails. If he selects heads, player 2 chooses between heads or tails. However, if he picks tails, she decides among heads, tails, or pain. It is obvious player 2 would never select pain as her strategy; it generates an extremely negative payoff for her. Yet the mere presence of it means she can infer player 1's original decision.

To see why, suppose player 1 picked heads:

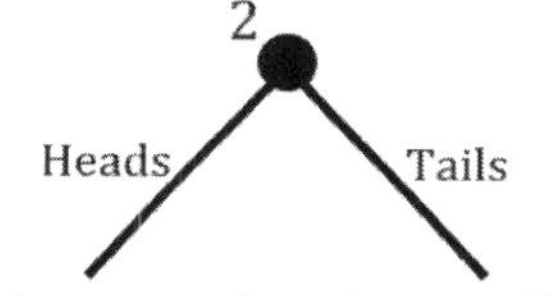

Player 2 must decide between heads or tails now, though she is not supposed to know which strategy player 1 chose. Nevertheless, she can infer that he selected heads. Why? Player 2 cannot play pain. The only way she could play pain is if player 1 selected tails. Since the pain strategy is not available, she therefore must be on the side of the game tree where player 1 played heads.

She can also infer when player 1 picked tails:

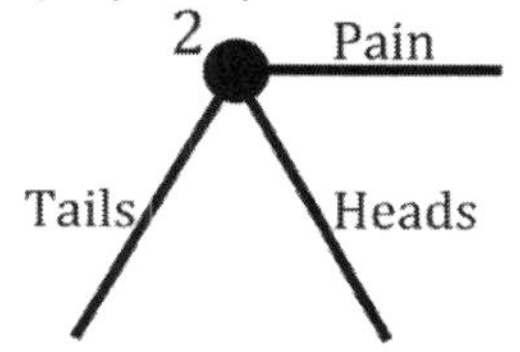

Now player 2 has the pain option available, so she immediately understands that she is on the tails side of the game. Either way, the simultaneous nature of the game is ruined. Thus, at any information set, a player must have the same strategies available regardless of how the player arrived there.

Second, player order is irrelevant in simultaneous move games. To see this, let's flip the order of moves around so that player 2 moves first:

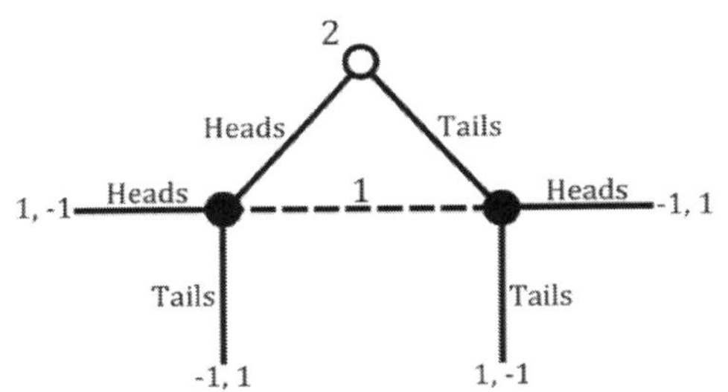

For the sake of consistency, player 1's payoff is still listed before player 2's, even though player 2 now moves first. Regardless, this game still converts to the same matrix as when player 1 went first:

	Left	Right
Up	1, -1	-1, 1
Down	-1, 1	1, -1

Obviously, this matrix is identical to the last one, so its Nash equilibrium is the same as well.

This has an important implication. Imagine we had to draw out the extensive form of the following games:

1) Player 1 picks heads or tails. Ten minutes later, player 2 picks heads or tails. Player 2 does not see player 1's selection when she decides.
2) Simultaneously, player 1 and player 2 pick heads or tails.
3) Player 2 picks heads or tails. Ten minutes later, player 1 picks heads or tails. Player 1 does not see player 2's selection when he decides.

Game theoretically, these are *identical* strategic situations, so we can use the same extensive form for all of them. Thus, even though player 2 picks heads or tails first in situation (3), we can draw the extensive form with player 1 moving first!

2.1.4: Why We Like Game Trees

When we want to analyze a strategic situation, knowing its extensive form is better than knowing its matrix. As it turns out, there is only one way to represent an extensive form game as a matrix. However, there can be multiple ways to represent a matrix in extensive form. Thus, if we only have the matrix in front of us, we do not know which of its Nash equilibria will survive subgame perfection.

Since we have seen simultaneous moves in extensive form games, we can see how two different game trees convert to the same matrix. Consider this tree:

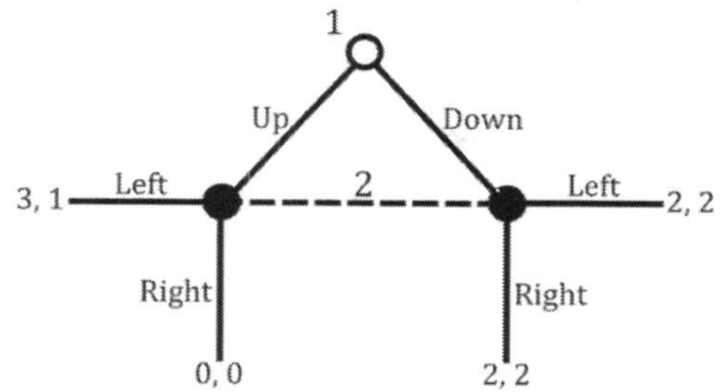

Note that the following matrix matches that extensive form game's moves and payoffs:

	Left	Right
Up	3, 1	0, 0
Down	2, 2	2, 2

Yet we used this exact matrix to describe Selten's game at the beginning of this lesson. So if we only had the matrix in front of us, we would not know whether the players move at the same time or if player 1 moved first, player 2 saw player 1's move, and then player 2 chose left or right. We know this has non-trivial consequences; <up, left> is the only subgame perfect equilibrium, but <up, left> and <down, right> are both Nash equilibria in the simultaneous move game.

Having the game tree in front of us eliminates any confusion from the start, which is why having the extensive form of a game is better. If we really need the matrix, the extensive form contains all the relevant information already. The conversion process does not work the other way around, however.

<u>Takeaway Points</u>

1) In sequential games, players take turns moving.
2) Subgame perfect equilibrium is the solution concept for extensive form games.
3) All subgame perfect equilibria are Nash equilibria, but not all Nash equilibria are subgame perfect equilibria.

Lesson 2.2: Backward Induction

Consider the "escalation game" below:

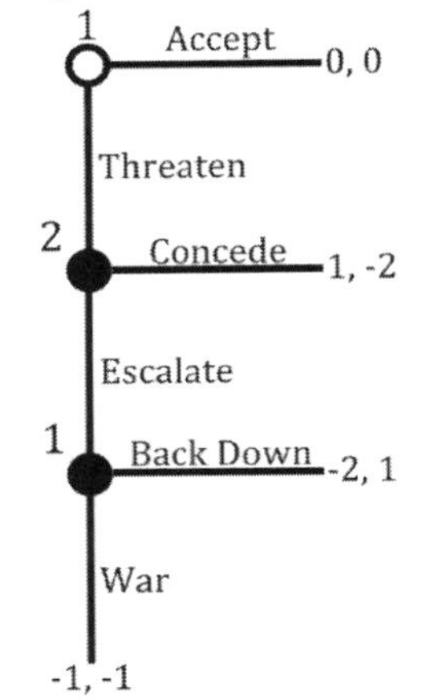

In this game, two countries are on the brink of war. Player 1 begins by accepting the status quo or issuing a threat. If he accepts the status quo, the game ends. If he threatens, player 2 decides whether to concede or escalate the conflict. The game ends if she concedes. If she escalates, player 1 chooses whether to launch war or back down. Either way, the game ends.

If player 1 accepts the status quo, each player earns 0. If player 2 concedes, player 1 makes a slight gain. Meanwhile, player 2 receives a slight loss and suffers a diminished reputation from the concession. Thus, player 1 earns 1 for this outcome, while player 2 earns -2. If player 2 escalates and player 1 backs down, the situation is reversed, and player 1 earns -2 while player 2 earns 1. Finally, if player 1 ultimately declares war, both sides suffer losses but save their reputations, giving both a payoff of -1.

Since this is an extensive form game, we must find its subgame perfect equilibrium. We could do this in two ways. First, we could convert the extensive form game into a matrix, find the game's Nash equilibria, and then work through the logic of the game tree to see if any of those Nash equilibria rely on incredible threats.

Alternatively, we could apply backward induction, which the easiest way to solve extensive form games when there are no simultaneous moves. Since the escalation game has no simultaneous moves, we will opt for backward induction here.

So what is backward induction? As the name suggests, when we use backward induction, we start at the end of the game and work our way to the beginning. Specifically, we see what the players would want to do at the end of the game and take that information to the previous step to see how players should rationally respond to those future moves. After all, the smartest move today depends on what will happen tomorrow. We repeat this process until we arrive at the beginning of the game.

Although that may sound complicated, backward induction is straightforward in practice. Let's use the escalation game to illustrate the process. We begin at the end of the game, when player 1 decides between war and backing down after player 2 escalates the conflict:

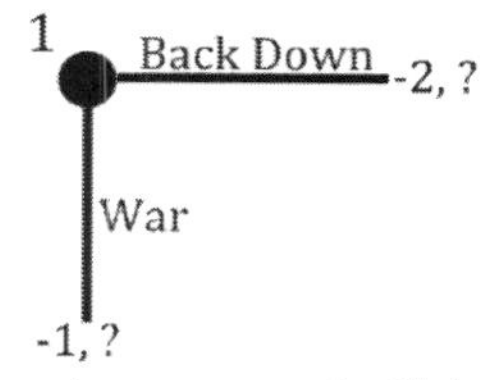

If player 1 declares war, he earns -1. If he backs down, he earns -2. Since -1 is greater than -2, we know player 1 declares war if he has the opportunity.

Now consider player 2's decision between conceding and escalating. She knows that player 1 will declare war if she escalates. As such, she can functionally ignore the outcome where player 1 backs down, as she knows that he will never play that strategy. Consequently, she can focus her decision between the following two outcomes:

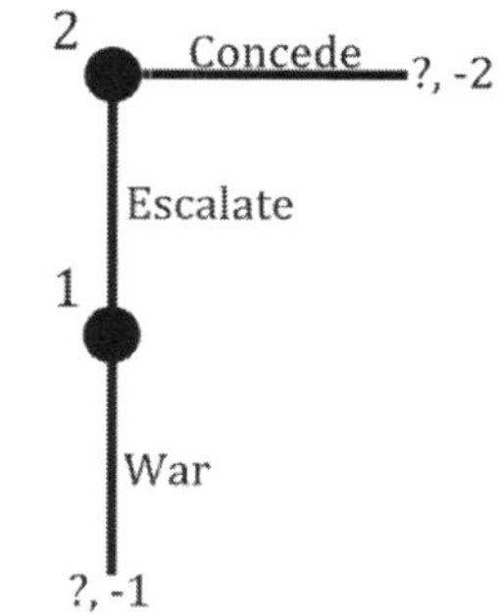

Essentially, we erased the back down outcome from the game. This allows us to concentrate on player 2's decision between conceding and escalating, knowing that player 1 will follow with war. If she concedes, she earns -2. If she escalates, player 1 declares war, and player 2 earns -1. Since -1 beats -2, player 2 escalates.

Knowing that, we can move to the beginning of the game where player 1 chooses whether to accept the status quo or issue a threat. We know that if he threatens player 2, she responds by escalating, which causes him to declare war. As such, player 1's choices can lead to the following outcomes:

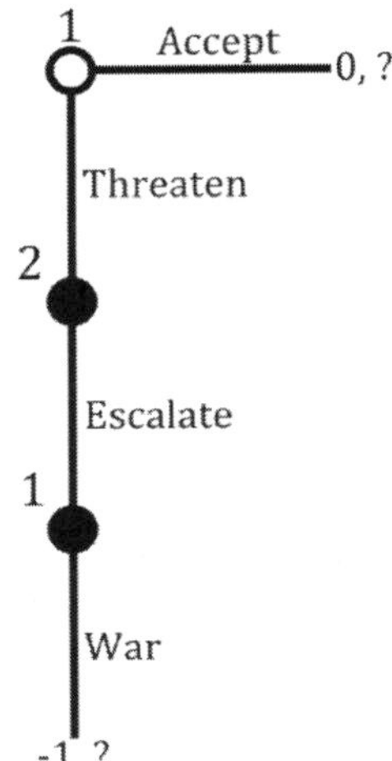

If he accepts the status quo, he earns 0. If he threatens player 2, she escalates, and he declares war, ultimately earning him -1. Since 0 is greater than -1, player 1 accepts the status quo and the game ends.

2.2.1: How Not to Write a Subgame Perfect Equilibrium

Differentiating between the outcome of an extensive form game and its subgame perfect equilibrium is extremely important. The outcome is what actually occurs when the players work their way through the game. The SPE is a complete and contingent plan of action for all players.

You might be tempted to say the SPE of the escalation game is for player 1 to accept the status quo. This is the outcome of the game. It is *not* the subgame perfect equilibrium. Why not? Subgame perfection, at its core, is the study of credible threats. Consequently, we want to know which threats in the escalation game are credible and which are not. To say that "player 1 accepts the status quo is the SPE of the escalation game" tells us nothing about the credibility of player 2's threat to escalate or player 1's threat to declare war. It is also not a complete and contingent plan of action, as it does not inform us what would happen if player 1 had to choose between backing down and war. A SPE must tell us all of this information.

Put differently, player 1 accepts the status quo at his initial decision node *because* player 2 will escalate if he issues a threat and *because* this ultimately causes him to declare war. Since the status quo beats war, he would rather accept the status quo. As such, we say that <(accept, war), escalate> is the SPE. This tells us that player 1 chooses accept and war at his two decision nodes, while player 2 selects escalate at hers.

In contrast, merely saying that player 1 accepts the status quo does not tell us why this choice is rational for him. Since subgame perfection is the study of credible threats, we need to know that information. As such, the SPE must list the optimal move at all decision nodes regardless of whether the players actually reach those nodes when they play their equilibrium strategies.

On the other hand, saying that "in the SPE, player 1 accepts the status quo, and the game ends" is accurate. Notice that the statement is not expressing the SPE but rather describing the outcome. As long as we are describing outcomes, we do not need to know what would occur off the equilibrium path of play. But as soon as we start discussing what the equilibrium itself is, we need all of that information.

2.2.2: Practice with Backward Induction

Since backward induction works on any extensive form game that does not have a simultaneous move, we ought to practice a bit more with it before moving on to other topics. Let's use it to see what happens when we turn some of the matrix games from last chapter into sequential move games. Let's start with the stag hunt:

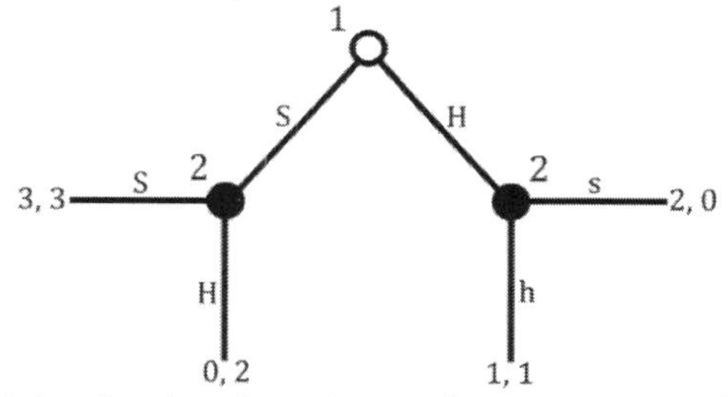

Here, player 1 begins by choosing to hunt a stag or hunt a hare. If he hunts a stag, player 2 sees this and chooses whether to hunt a stag or hunt a hare. Similarly, if he hunts a hare, player 2 sees this and chooses whether to hunt a stag or hunt a hare.

For this game, notice that we have differentiated player 2's moves depending on what player 1 chooses. Specifically, if player 1 played S, player 2 chooses between S and H. But if player 1 played H, player 2 chooses between s and h. Alternating between capital and lower case letters allows us to differentiate between these strategies, since S (hunting a stag if player 1 hunted a stag) is fundamentally different from s (hunting a stag when player 1 hunted a hare).

The stag hunt also includes a new obstacle. Selten's game and the escalation game both had a single final possible decision. Here, there are two possible final decisions, depending on what player 1 selected:

When we applied backward induction to the escalation game, we knew we had to start with player 1's decision to back down or declare war. How do we select between the two possible end games of the stag hunt?

Fortunately, the choice is irrelevant. We can start at *any* end of the game and work our way backward until we cannot go further. Then we switch to a different end of the game and work our way backward until we cannot go further. We repeat this as many times as necessary until we have solved the game.

Let's illustrate this process using the stag hunt. Since player 2's decision between s and h is a terminal node, let's start there:

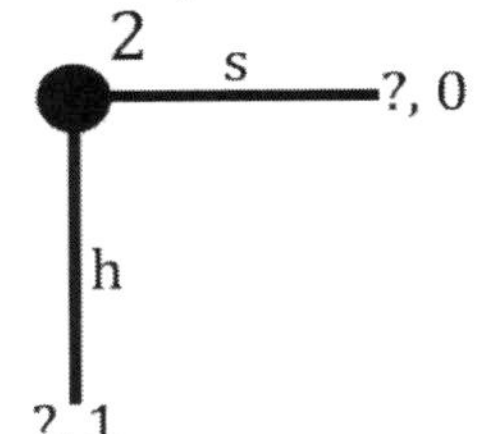

Player 2 knows player 1 is hunting a hare. If she selects s, she fails to capture the stag and earns 0. If she selects h, the players split the hares, and she earns 1. Since 1 beats 0, she chooses h. If we erase s from the game, we end up with this:

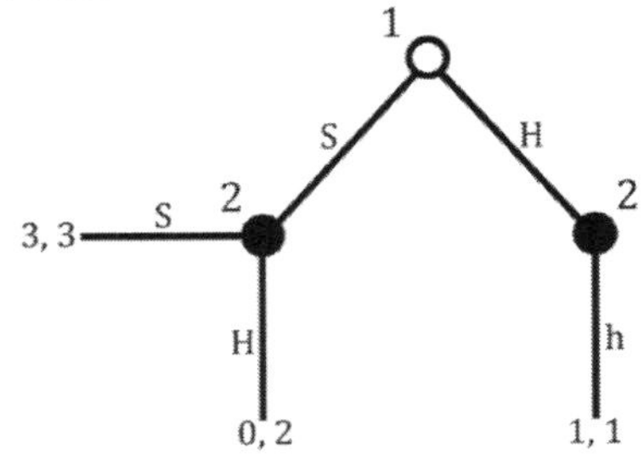

With the escalation game, we could just go to the previous decision node and decipher that player's optimal decision. However, if we tried solving for player 1's move here, we would run into a problem. If he selects H, we know that player 2 plays h as well. However, if he chooses S instead, we do not yet know whether player 2 will hunt a stag or a hare. As such, we have to work our way back up from that terminal node before we can solve for player 1's strategy.

So suppose player 1 hunts a stag. Player 2 faces the following decision:

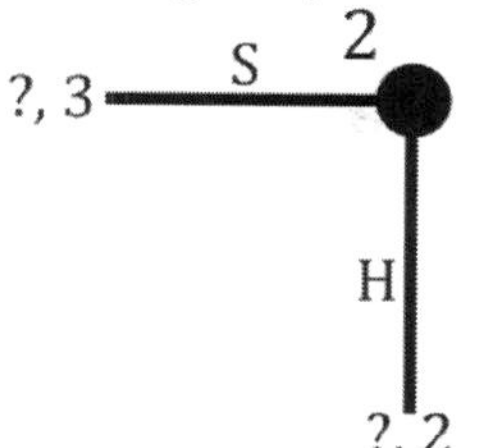

Player 2 knows player 1 is hunting a stag. If she selects S, they capture the stag, and she earns 3. If she selects H, she earns 2. Since 3 beats 2, she chooses S. If we erase H from the game, we end up with this:

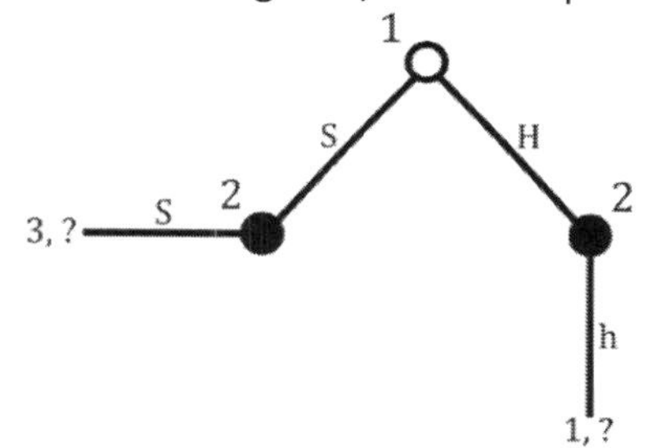

Now player 1 knows exactly what player 2 does after he moves. If he plays S, she moves S, and he earns 3. On the other hand, if he chooses H, she selects h as well, and he earns 1. Since 3 beats 1, player 1 plays S.

Therefore, the game's SPE is <S, (Sh)>; in the SPE, both players hunt a stag. Effectively, the sequential nature of the game solves the coordination problem. In the original stag hunt, <hare, hare> was a Nash equilibrium. But since player 1 can establish that he is hunting a stag, player 2 never has a reason to play hare in response.

Sequentiality also resolves the coordination problem in battle of the sexes:

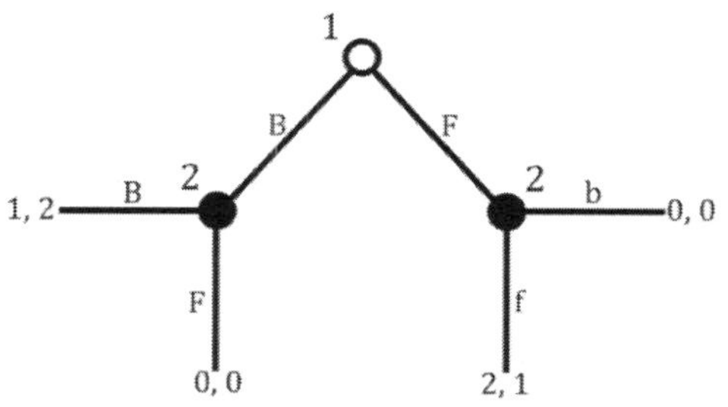

Let's start at player 2's decision between ballet and fight after player 1 has gone to the ballet:

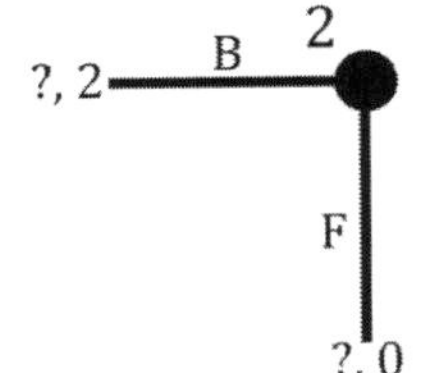

Since 2 beats 0, player 2's optimal strategy is to play B. Erasing F yields the following:

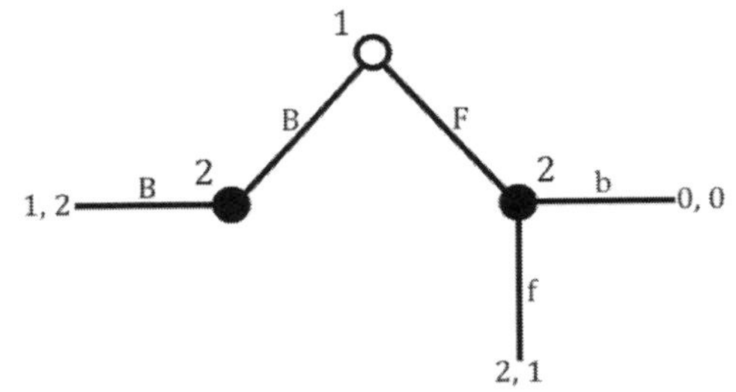

We cannot yet solve for player 1's optimal strategy since we do not yet know how player 2 will respond if he goes to the fight. So let's analyze that contingency:

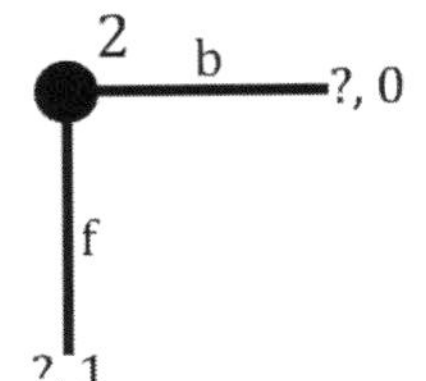

Here, she earns 1 for f and 0 for b. As such, she chooses f. If we erase b, we are left with the following:

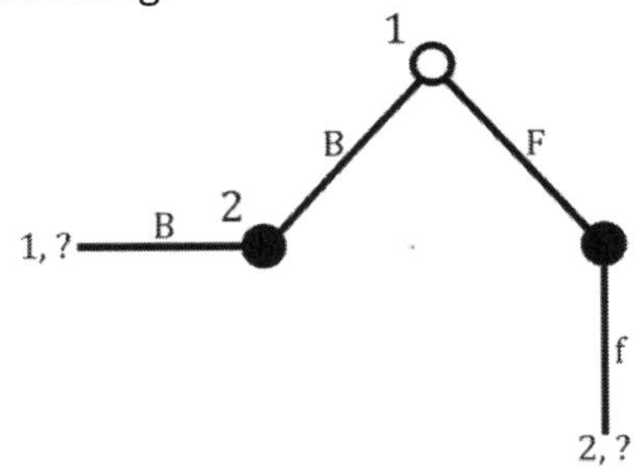

Now player 1 knows if he goes to the fight, she goes there as well, and he earns 2. In contrast, if he elects to see the ballet, she heads there too, and he only earns 1. Since 2 beats 1, he goes to the fight.

Thus, the game's SPE is <F, (Bf)>; in the SPE, both players meet at the fight. Unlike in the simultaneous move battle of the sexes, this version of the game has a single equilibrium. Again, the sequentiality of the game has resolved the coordination problem between the players.

Indeed, player 1 actually exploits his first mover's advantage. In simultaneous move game, <ballet, ballet> was also a Nash equilibrium, meaning player 2 could arrive at her most preferred outcome. Here, that is not possible. Player 1 simply goes to his most preferred form of entertainment, knowing that player 2 must follow.

First movers do not always have an advantage, though. Consider the weighed game of matching pennies that we used to introduce the mixed strategy algorithm:

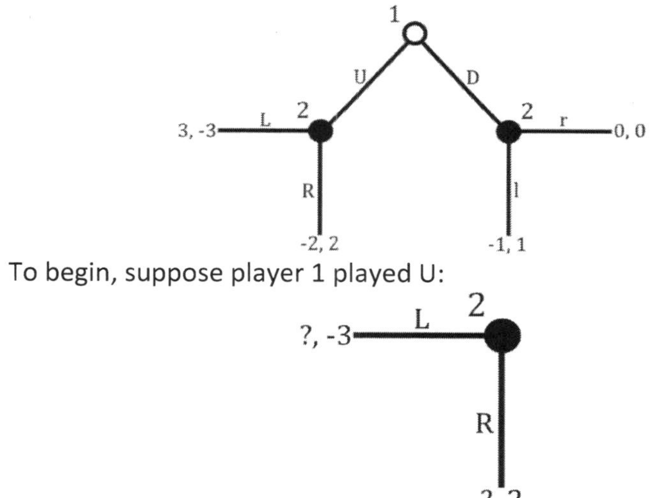

To begin, suppose player 1 played U:

Player 2 can choose L and earn -3 or play R and earn 2. Naturally, she will play R in this contingency.

Let's switch over to the other terminal node and suppose player 1 played D:

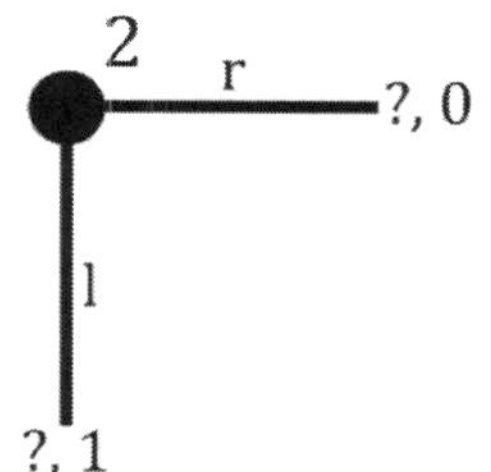

This time, she earns 1 for l and 0 for r. As such, she picks l.

Consequently, player 1's decision is between the following:

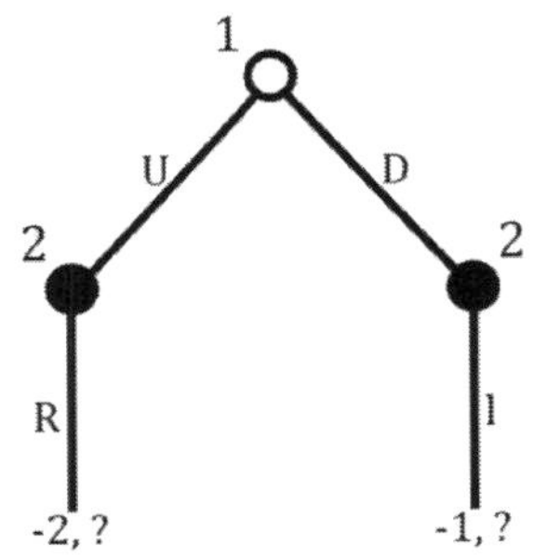

Up is worth -2, while down nets him -1. Consequently, he selects down. Since player 2 moves last, she can always counteract player 1's move. Knowing this, all he can do is minimize his losses. Here, that means taking down and -1 over up and -2. Therefore, the SPE is <D, (Rl)>.

While you absolutely must know how to use backward induction if you are serious about game theory, do not worry if you are not yet comfortable with it. In the remaining lessons, we will look at plenty of games that we can solve with backward induction.

Takeaway Points

1) Backward induction finds subgame perfect equilibria by finding how players optimally behave at the end of the game and uses that information to find how players optimally behave at the beginning of the game.
2) A subgame perfect equilibrium is a complete and contingent plan of action. It must state what all players would do at a particular decision node regardless of whether they actually reach that node in equilibrium.

Lesson 2.3: Multiple Subgame Perfect Equilibria

One of the reasons we like using backward induction is that it generally produces a unique subgame perfect equilibrium. However, some games break that rule. This section covers how these exceptions come up and why they can lead to problems.

2.3.1: The Ultimatum Game

Let's start with a simple example. Player 1 has some good worth a value of 2 and has to bargain with player 2 over how to divide it. He can offer to split the good or he can attempt to take all of it. However, player 2 can reject either proposal. If she does, both receive nothing.

Here is the game tree:

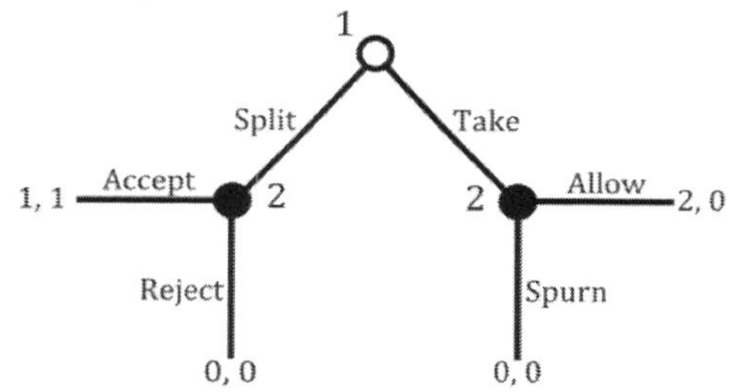

This is a simple version of the ultimatum game. Player 1 begins by making an ultimatum—split or take. If he splits, player 2 accepts or rejects his division. If she accepts, they both earn 1; otherwise, they both earn 0. On the other side, if he takes, player 2 allows that to happen or spurns player 1's move. If she allows it, player 1 earns 2 while she earns 0. But if she spurns him, they both earn 0.

There are no simultaneous moves here, so let's use backward induction. First, consider player 2's choice if player 1 splits:

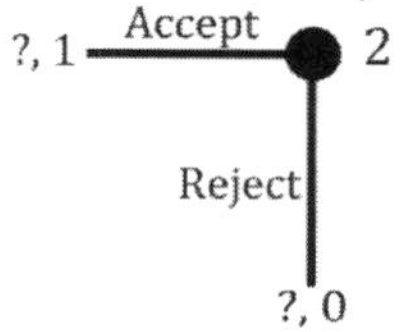

If player 2 accepts, she earns 1. If she rejects, she earns 0. Since 1 beats 0, player 2 always accepts if player 1 splits.

Now let's go to the other side of the game tree:

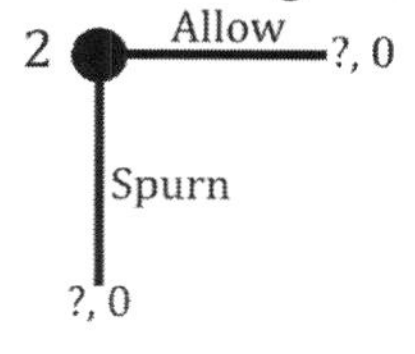

Regardless of what player 2 does here, she earns 0. Thus, in equilibrium, she can do three things: play spurn as a pure strategy, play allow as a pure strategy, or mix between allow and spurn according to any probability distribution. All of these choices generate the same payoff and are therefore optimal for her in this subgame.

Thus, we have to consider all three cases when we move backward a step. Let's begin by assuming she always allows:

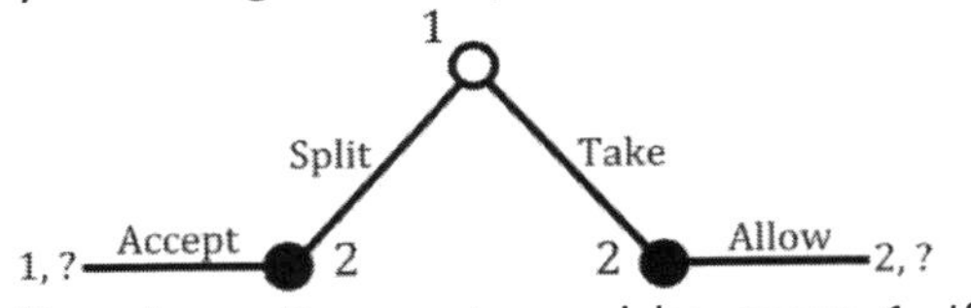

If player 1 splits, player 2 accepts, and he earns 1. If player 1 takes, player 2 allows, and he earns 2. Since 2 beats 1, player 1 takes. So if player 2 always allows after player 1 takes, the subgame perfect equilibrium is <take, (accept, allow)>.

But notice what happens if player 2 always spurned instead:

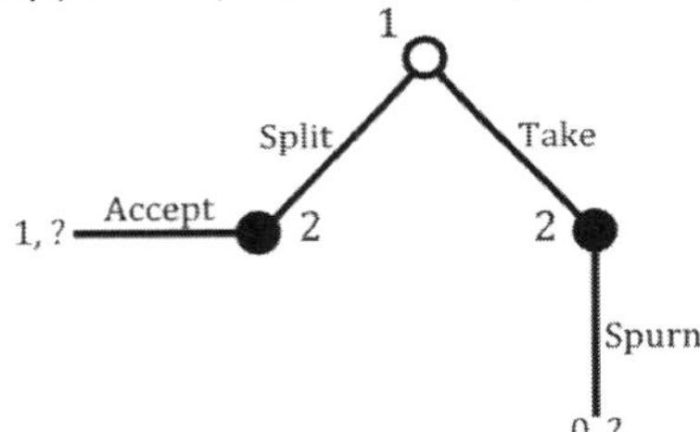

Player 1 still ultimately earns 1 if he splits. However, if he takes, player 2 spurns, and he earns 0. Since 1 beats 0, player 1 splits. Thus, if player 2 always spurns after player 1 takes, the subgame perfect equilibrium is <split, (accept, spurn)>.

So far, we have found two distinctly different subgame perfect equilibria. But it gets worse once we consider what happens when player 2 mixes between allow and spurn:

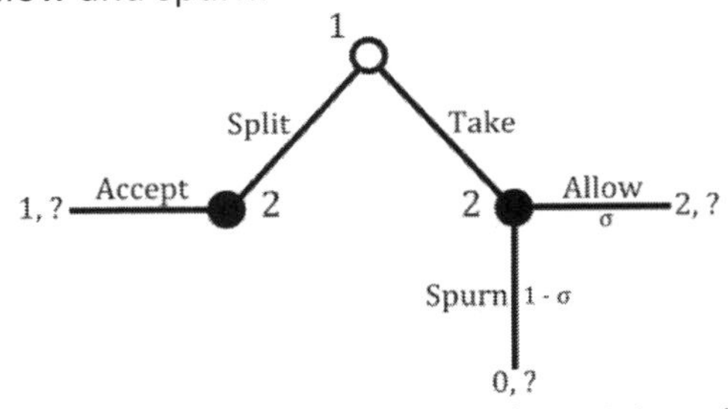

Once again, player 2 accepts if player 1 splits, giving player 1 a payoff of 1. If he takes, player 2 allows with probability σ_{allow} and spurns with probability $1 - \sigma_{allow}$. When she allows, he earns 2; when she spurns, he earns 0. Thus, his expected utility for taking is:

$$EU_{take} = (\sigma_{allow})(2) + (1 - \sigma_{allow})(0)$$
$$EU_{take} = 2\sigma_{allow}$$

Player 1's expected utility for playing split is 1. Therefore, he must play take in SPE if $2\sigma_{allow}$ is greater than 1:

$$EU_{split} = 1$$
$$EU_{take} = 2\sigma_{allow}$$
$$EU_{take} > EU_{split}$$
$$2\sigma_{allow} > 1$$
$$\sigma_{allow} > 1/2$$

Thus, if player 2 allows more than half of the time, player 1 must always take, as his expected utility for taking is always greater than 1 in that instance. As such, for any $\sigma_{allow} > 1/2$, the SPE is <take, (accept, $\sigma_{allow} > 1/2$)>.

But there are also situations in which player 2 mixes and player 1 must optimally split:

$$EU_{split} = 1$$
$$EU_{take} = 2\sigma_{allow}$$
$$EU_{take} < EU_{split}$$
$$2\sigma_{allow} < 1$$
$$\sigma_{allow} < 1/2$$

Thus, if player 2 allows less than half of the time, player 1 must always split, as his expected utility for taking is always less than 1 in that instance. As such, for any $\sigma_{allow} < 1/2$, the SPE is <split, (accept, $\sigma_{allow} < 1/2$)>.

Finally, note that if player 2's mixed strategy is exactly $\sigma_{allow} = 1/2$, player 1 is indifferent between splitting and taking:

$$EU_{split} = 1$$
$$EU_{take} = 2\sigma_{allow}$$
$$EU_{take} = EU_{split}$$
$$2\sigma_{allow} = 1$$
$$\sigma_{allow} = 1/2$$

That being the case, player 1 can mix according to any probability distribution between splitting and taking; regardless of what he does, his expected utility equals 1. Thus, for σ_{allow} = 1/2, the SPE is <σ_{split}, (accept, σ_{allow} = 1/2)>, where σ_{split} can be any probability.

On substantive grounds, any SPE in which player 2 allows player 1 to take seems implausible. Because player 2 is indifferent between allowing and spurning, she has a credible threat to spurn. Note that if she plays spurn in an SPE, player 1 must split at the beginning of the game. After he splits, she accepts, and she walks away with 1 as her payoff, which is the best outcome she could possibly receive. Consequently, it would be bizarre if player 2 did not force player 1 to split. Yet the aforementioned cases in which player 2 allows with positive probability are still technically subgame perfect equilibria.

The reason this game has multiple equilibria is that player 2's payoffs are the same for each of her choices at one of the decision nodes:

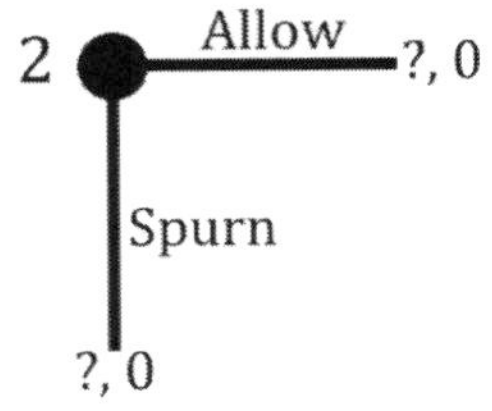

This allows player 2 to choose either pure strategy or mix in any way between them. Absent that, she would have to pick a single strategy with certainty, thereby bringing us back to a unique SPE.

2.3.2: Multiple Equilibria, Same Outcome

The last example had different outcomes depending on which subgame perfect equilibrium the players selected. However, there are trivial cases with a unique equilibrium outcome to the game but multiple subgame perfect equilibria. Here is an example:

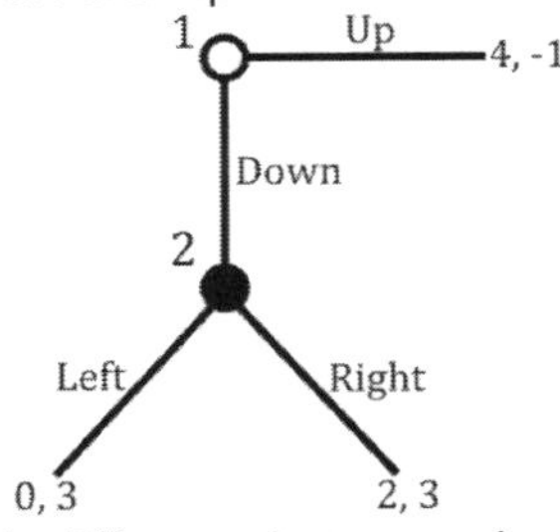

Note that player 2 is indifferent between her two strategies if player 1 moves down:

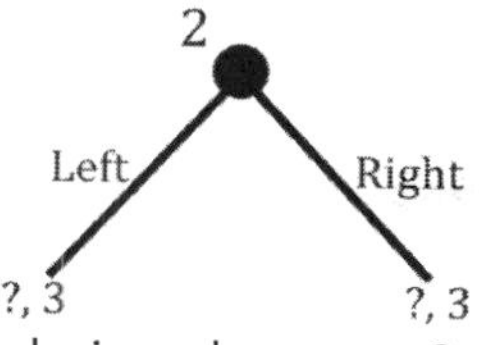

Regardless of player 2's choice, she earns 3. Therefore, she can play left as a pure strategy, play right as a pure strategy, or mix freely between left and right.

However, player 1 will never let player 2 make a move:

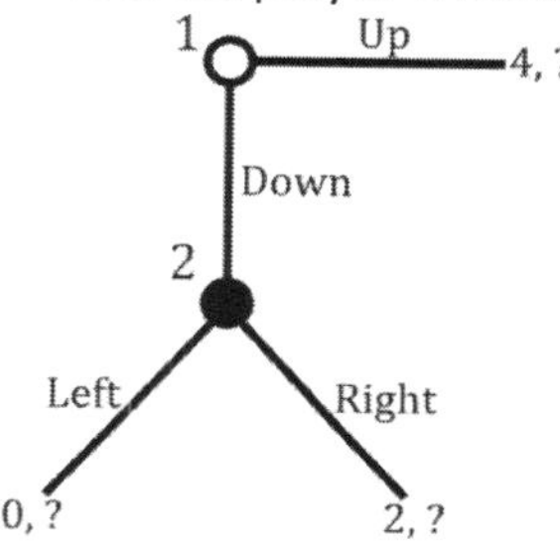

If player 1 moves down, then the best possible payoff he can finish with is 2. In contrast, if he plays up, he earns 4. Thus, in the SPE, player 1 plays up and player 2 adopts any strategy she wishes.

2.3.3: When There Must Be a Unique SPE

In a sequential game with no simultaneous moves, if an individual's payoffs are different for every outcome, and this is true for all individuals in the game, backward induction *must* yield a unique solution. For example, suppose a game has three possible outcomes, player 1's payoffs for these outcomes are -1, 2, and 5, and player 2's payoffs for these outcomes are 2, 3, and 9. -1, 2, and 5 are all distinct numbers. Likewise, so are 2, 3, and 9. Therefore, at every individual decision node, each player has a single optimal decision, and the SPE will be unique.

As this example also shows, it does not matter if the players have a common payoff. In this case, both players can possibly earn 2. Nevertheless, multiple equilibria are not possible, as that would require player 1 to have two separate outcomes that yield him 2 or player 2 to have two separate outcomes that yield her 2. As long as the duplicate payoff is spread around to different players, the SPE is unique.

Why is this the case? For a player to be able to choose optimally among more than one strategy, his expected utility must be the same for all of those outcomes. However, if his payoffs are all different, then he cannot be indifferent. Exactly one of those choices must be optimal for him, and he must make that choice in equilibrium.

2.3.4: Multiple Equilibria with Simultaneous Moves

One caveat to this rule is that we must use backward induction on the game, so the game must be sequential all the way through. If a single simultaneous decision exists in the game tree, multiple SPE may exist even if each payoff is unique. Here is an example:

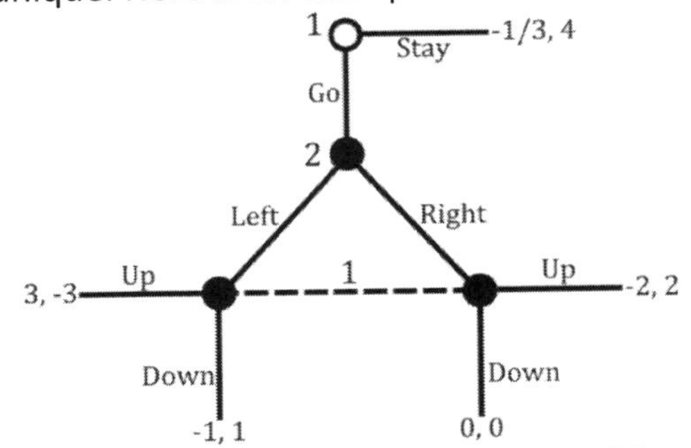

Player 1 begins the game by staying or going. If he goes, the players face off in the weighted matching pennies game that we used to introduce the mixed strategy algorithm. If he stays, the game ends and he earns -1/3 and she earns 4.

This game also presents a new problem in that we cannot use backward induction here. Backward induction requires every decision node to have a unique history. Here, however, when player 1 reaches his information set where he chooses between up and down, he does not know whether player 2 selected left or right. So we cannot start at the bottom and work our way up.

Fortunately, the solution is to utilize the *subgame* part of subgame perfect equilibrium. Rather than working from the very bottom, we work from the last decision in the game with a unique history. We call this a subgame.

Note this means we cannot use player 1's last move as the start of the subgame:

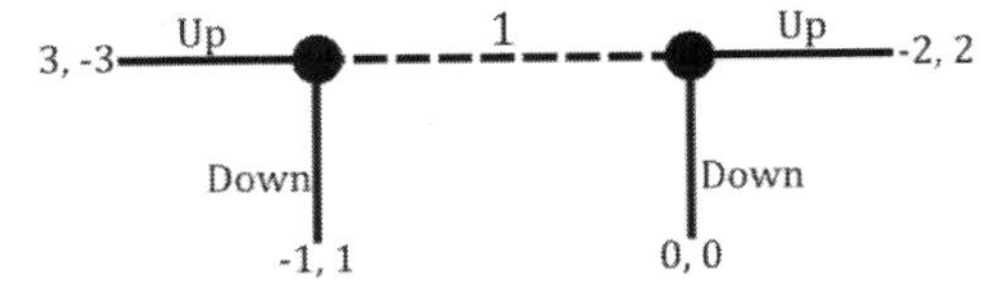

This is an invalid subgame. Player 1 does not know exactly where he is—player 2 played left or right, but player 1 does not directly observe which. Moreover, if player 2 moved left, player 1 should move up; yet if player 2 went right, player 1 should go down. Backward induction cannot help us here.

However, when player 2 chooses left or right, she knows exactly what has happened before—namely, player 1 chose go. So we begin by analyzing that subgame:

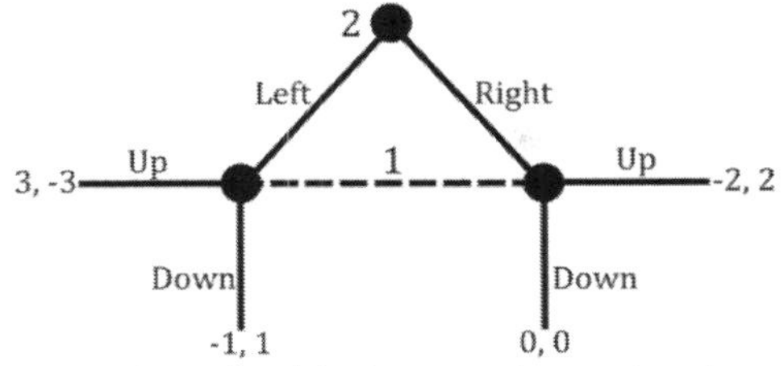

As previously mentioned, this is merely a simultaneous move game identical to the weighed matching pennies game from earlier. Thus, to solve this subgame, we find the Nash equilibria of its matrix:

	Left	Right
Up	3, -3	-2, 2
Down	-1, 1	0, 0

From Lesson 1.5, we know that player 1 mixes, playing up with probability 1/6 and down with probability 5/6. Player 2 also mixes, playing left with probability 1/3 and right with probability 2/3. In the mixed strategy Nash equilibrium, player 1's expected utility equals -1/3, while player 2's equals 1/3.

With that information in hand, we erase the subgame and replace it with those payoffs:

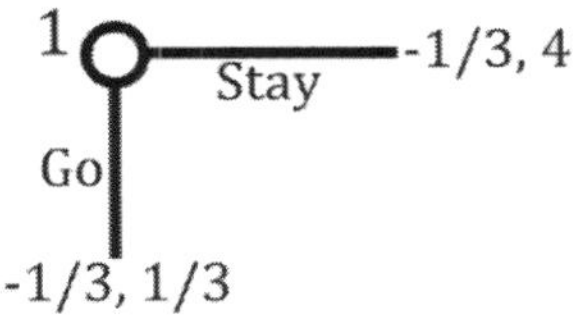

The remaining game is similar to the other extensive form games we have seen in the past. At this point, we merely find player 1's optimal strategy. As it turns out, he earns -1/3 regardless of whether he chooses stay or go. As such, he can select either as a pure strategy or play any mixture between the two. Thus, we must write out three different types of SPE. In the first, player 1 goes as a pure strategy, giving us an SPE of <(Go, σ_{up} = 1/6), σ_{left} = 1/3>. In the second, player 1 stays as a pure strategy. Although the players never play the weighed matching pennies subgame, we still must list their mixtures for that subgame in the SPE. Therefore, we write that SPE as <(Stay, σ_{up} = 1/6), σ_{left} = 1/3>. Finally, <(σ_{go} , σ_{up} = 1/6),

σ_{left} = 1/3> represents the cases where player 1 mixes between stay and go, where σ_{go} equals any number between 0 and 1, not including 0 and 1.

Notice multiple subgame perfect equilibria exist despite how each player's payoffs are unique for that player. Specifically, player 1 can earn -2, -1, -1/3, 0, and 2 depending on the outcome, while player 2 can earn -3, 0, 1, 2, and 4 depending on the outcome. However, player 1's expected utility of the simultaneous move subgame equals his payoff for the outcome outside of that subgame, which leads to the indifference that allows for multiple equilibria.

Not all extensive form games with at least one simultaneous move have multiple SPE. Indeed, the previous game required player 1's payoff for the stay outcome to be *exactly* -1/3 for multiple equilibria to exist. Had his payoff for staying been anything higher, his optimal strategy would be to stay, as playing the weighed matching pennies game has an expected utility of -1/3. Similarly, had the payoff for staying been anything lower, his optimal strategy would be to play the weighed matching pennies game.

Takeaway Points

1) If each player's payoffs are unique for that player and the game has simultaneous moves, the game has a unique SPE.
2) If a player has duplicate payoffs or the game has a simultaneous move, the game could have multiple SPE.

Lesson 2.4: Making Threats Credible

We often think that keeping our options open is the most prudent course of action. After all, if you limit your future choices, you might not be able to optimally respond to competing behavior.

However, that theory falls flat when we think about credible threats. This lesson shows why players might want to intentionally constrain their future actions. There are two different ways to accomplish this. First, players can burn bridges—that is, make a certain future course of action impossible. Second, they can tie their hands—that is, leave a future option open but make it so extremely undesirable that they would never choose to pursue it. Regardless of the specific method, burning bridges and tying hands allows players to make their threats credible and in turn increase their equilibrium payoffs.

2.4.1: Burning Bridges

The bridge-burning story is the classic example. Although Thomas Schelling most frequently receives attribution for this tale, similar parables probably predate written history.

The story goes as follows. Two countries are at war. A small island sits between the two. Each country has only one bridge that can access it. Although valuable, the island is not worth fighting over; each side would rather concede the territory to its opponent than fight a battle over the land.

The first country crosses its bridge to occupy the island. Afterward, the soldiers decide whether to burn the bridge behind them. The second country decides whether to invade. If the first country has no bridge to use as an escape route, it must fight a battle. However, if the bridge still stands, the first country decides whether to fight or retreat.

Here is the game tree:

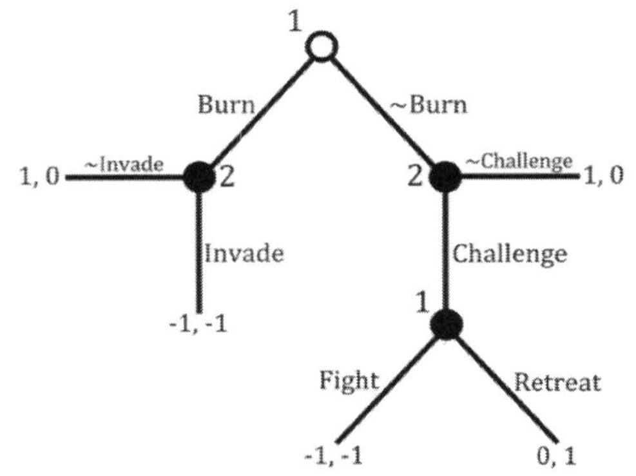

We can solve this game using backward induction. Let's begin with player 1's decision to fight or retreat:

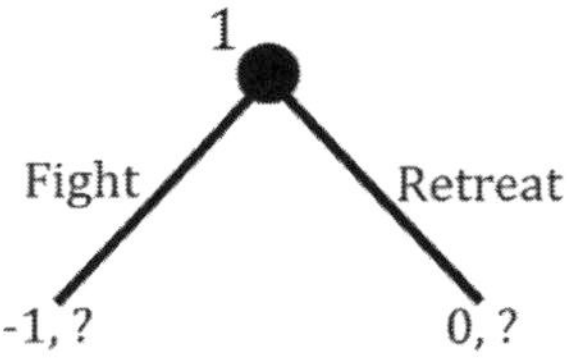

Since 0 beats -1, player 1 retreats. This option is only available to player 1 because he did not burn the bridge at the start of the game.

Let's move back to player 2's decision whether to challenge player 1 for control over the island:

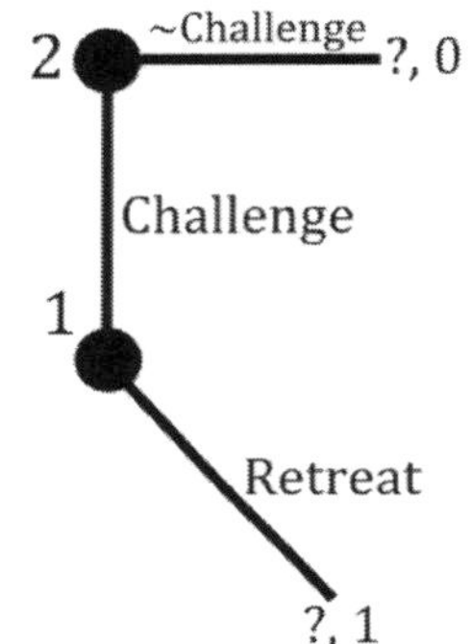

Player 2 sees that player 1 left the bridge unburned, so she knows he retreats if challenged. Eventual retreat from player 1 generates a payoff of 1 for player 2. Meanwhile, challenging gives her 0. Since 1 beats than 0, player 2 challenges if player 1 does not burn the bridge.

Now let's switch to the side of the tree where player 1 burns the bridge:

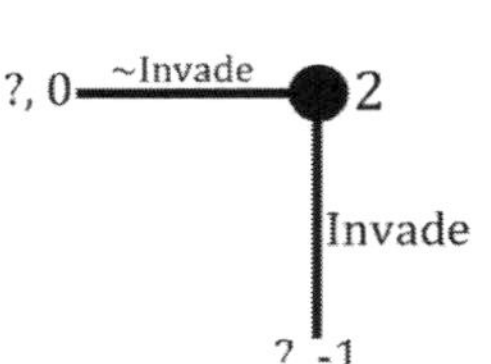

If she invades, player 1 cannot retreat since he burned his escape route, and thus the players end up in a battle. Player 2 earns -1 for that outcome. Alternatively, she could not invade the island and concede control of it to player 1. She earns 0 for that outcome. Since 0 beats -1, she opts not to invade.

We can now analyze player 1's initial decision whether to burn the bridge:

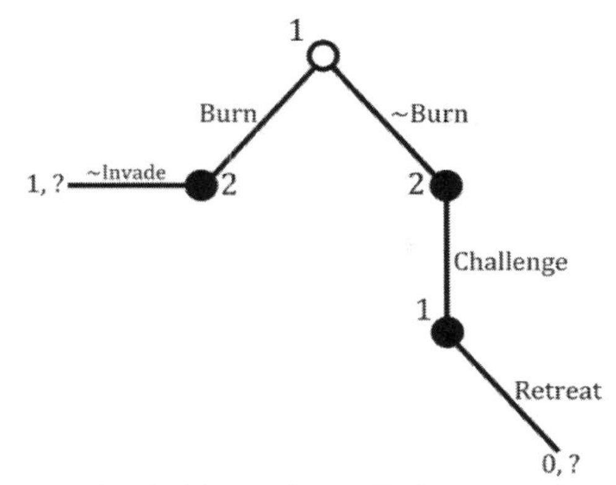

If player 1 burns the bridge, player 2 does not invade, and player 1 earns a payoff of 1. If he does not burn it, player 2 challenges him for control over the island, which forces player 1 to retreat. Player 1 earns 0 for this outcome. Since 1 beats 0, player 1 optimally begins the game by burning the bridge.

Therefore, the SPE is <(burn, retreat), (not invade, challenge)>. Strikingly, player 1 can make an incredible threat credible by burning the bridge; he cannot credibly threaten to fight for control over the island, so he puts himself into a position where he has no choice but to do so if player 2 invades. Player 2 sees the credibility of this threat and opts not to walk into a battle on the island.

Absent the burned bridge, however, player 1 cannot commit to fighting over the island, which prompts player 2 to take control of it.

2.4.2: Tying Hands

A boss notices that one of his unscrupulous employees has been stealing company materials lately. He values honesty in himself and his employees, but the stolen property was not valuable. Consequently, the boss prefers keeping her around rather than having to hire and train a replacement. Nevertheless, he would ideally like stop her from stealing.

At the company meeting today, he is thinking about issuing a warning: the next person caught stealing any company property will be immediately fired. Should he issue such a warning?

Here is the game tree:

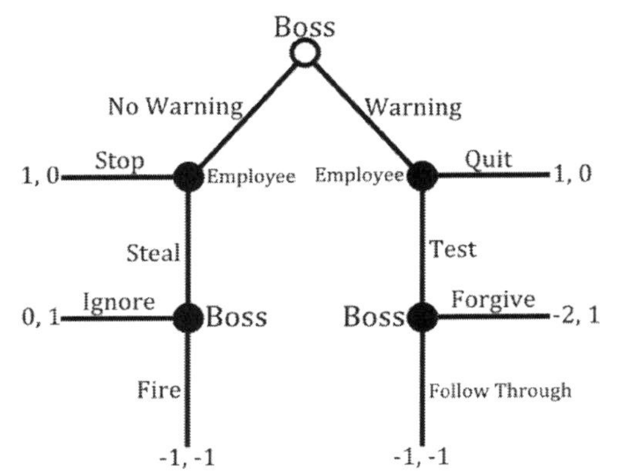

Let's begin the analysis on the left side. Suppose the boss issued no warning and an employee stole something. How does the boss respond?

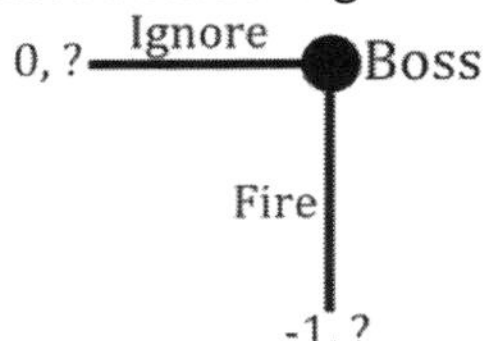

Ignoring pays 0; firing pays -1. Since the stolen good is not valuable, the boss ignores the theft in this situation.

Now let's move back to the employee's decision to steal or not:

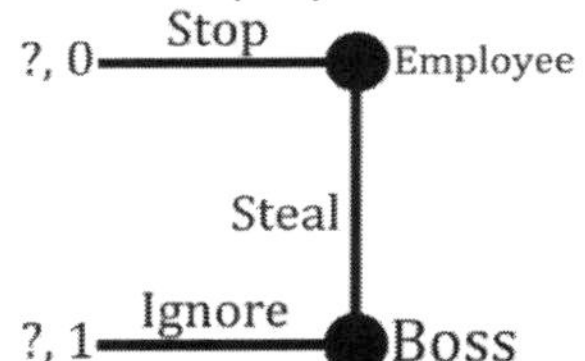

If the employee steals something, the boss ignores the theft, and she earns 1. If she stops stealing, she earns 0. Since 1 is greater than 0, she steals.

Let's switch to the other side of the game tree. Suppose the boss issued a warning and an employee stole something anyway:

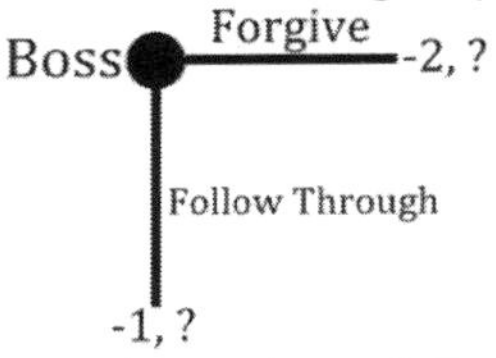

This time, the boss follows through and fires the employee. He values his honesty and reputation; if he does not fire the employee, he will develop a reputation of being a liar. Consequently, he earns -2 for ignoring the theft and -1 for firing the employee. Since -1 is greater than -2, he fires her.

Now we go back to the employee's decision to steal from the office if the boss warns her not to:

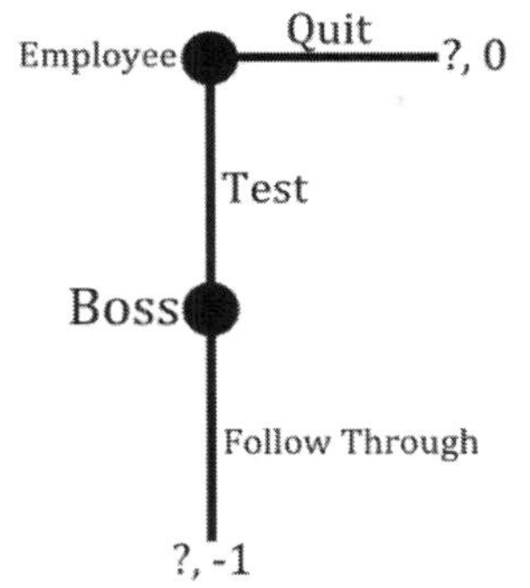

If the employee tests her boss's word, the boss fires her, and she earns -1. However, if she quits stealing, she earns 0. As such, she quits.

With that, we can now find the boss's optimal decision to begin the game:

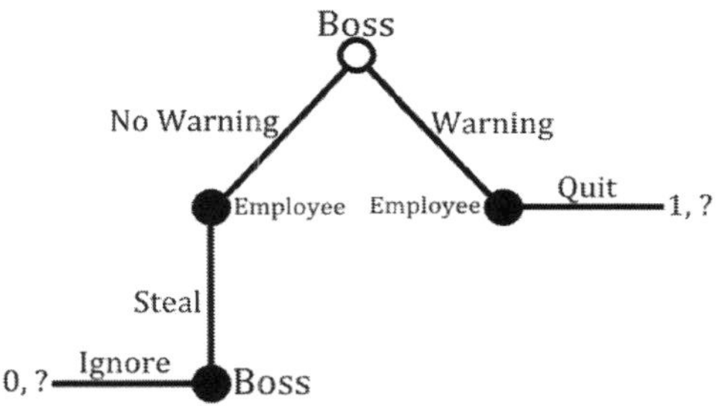

If the boss issues a warning, his employee quits stealing, and he earns 1. If he does not issue a warning, the employee steals, he ignores it, and he earns 0. Since 1 beats 0, he issues the warning.

Thus, the subgame perfect equilibrium is <(warning, ignore, follow through), (steal, quit)>. In the SPE, the boss warns his employee, and she does not steal again.

<u>Takeaway Points</u>

1) Subgame perfect equilibrium is the study of credible threats.
2) Sometimes players can gain by constraining their future actions to make threats credible. This deters the other player from taking aggressive actions they would have otherwise.

Lesson 2.5: Commitment Problems

Suppose you are a college graduate from San Diego, California, and you were recently admitted to a political science PhD program in Rochester, New York. Naturally, you pack up all of your earthly belongings into your compact Honda Civic, cover everything with an old sheet to shield the items from the prying eyes of a potential thief, and embark on a cross-country adventure to your new home.

But trouble strikes halfway there. As you are driving through El Paso, Texas, police lights flash behind you. You pull over and roll down your window. The officer explains that El Paso is in the middle of a drug war and that you appear suspicious, coming from California in a vehicle filled with unknown objects under a sheet. He politely requests to conduct a search of your vehicle.

You tell him you are a graduate student moving from San Diego to Rochester and object to such a search, noting that he has no legal right to look through your belongings.

Begrudgingly, the officer accepts that he cannot search your vehicle without permission. However, he notes that he could call in a K-9 unit to sniff around the vehicle. But the K-9 unit is stationed a half hour away, so it would take a while for it to arrive. He suggests a compromise: you allow him to conduct a quick search, and you can be on your way in a few minutes. He stresses that the quick search will be better than the K-9 for both parties, as neither of you will have to wait in the hot summer sun.

Should you take the officer's offer?

Having studied game theory, you mentally draw out the game tree. Your move is first: you can either demand the K-9 unit or allow the officer to search. If you allow the search, the officer decides between conducting a quick search as he originally offered or reneging on that agreement and conducting an extensive search. You most prefer a quick search and least prefer an invasive extensive search. Meanwhile, the police officer would most like to conduct an extensive search to ensure you are not carrying drugs but least prefers waiting a long time for the K-9 to arrive.

That game tree looks like this:

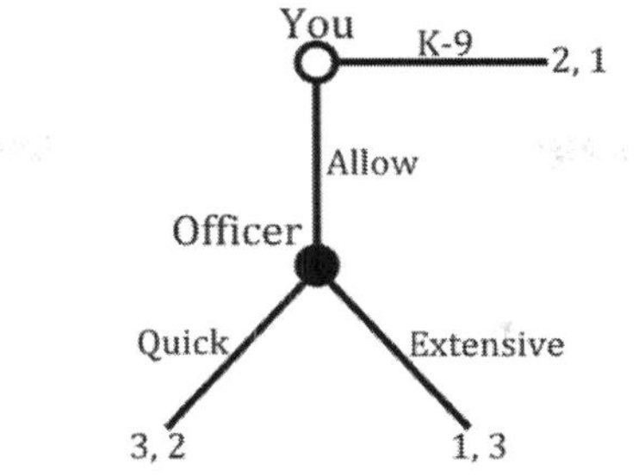

This is a simple sequential game of complete information. Backward induction swiftly solves it. Begin at the end of the game, when the officer decides what type of search to conduct:

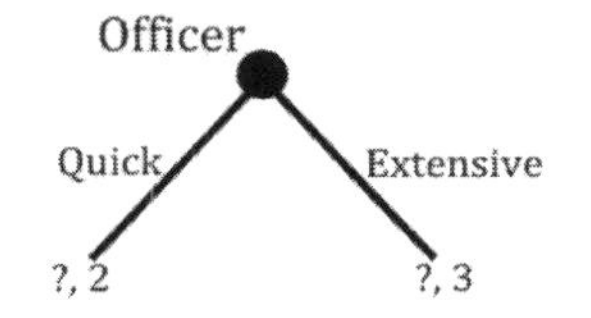

Since 3 beats 2, he chooses an extensive search. As such, we remove quick search from the game and see how you should play at the beginning, knowing that the officer will conduct an extensive search:

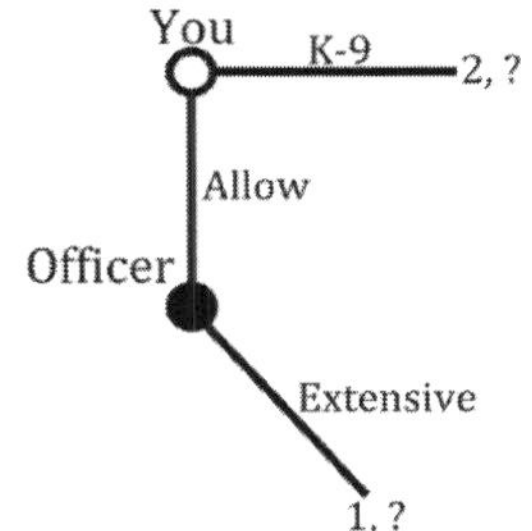

If you request the K-9, you earn 2. If you allow a search, the officer chooses an extensive search, and you earn 1. Since 2 beats 1, you choose K-9. Therefore, the SPE of this game is <K-9, Extensive>.

Objectively, this is an unfortunate outcome. Let's compare the K-9 outcome to the quick search outcome:

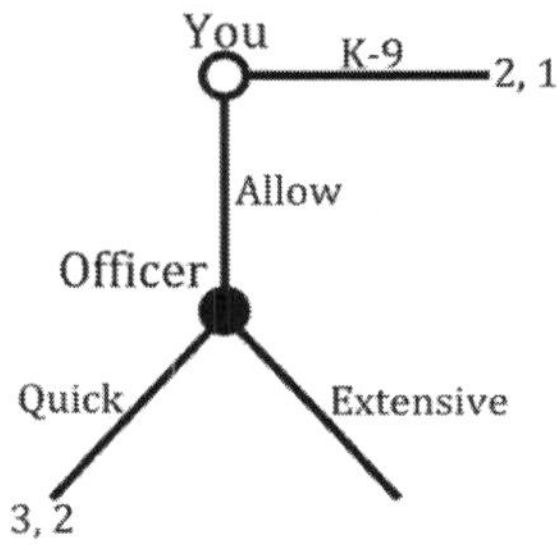

At the beginning of the story, the officer told you that a quick search would be better for both of you than waiting for the K-9 unit. He was right. A quick search is worth 3 for you, whereas waiting for the K-9 is worth 2. Meanwhile, a quick search is worth 2 for the officer, whereas waiting for the K-9 is worth only 1.

Unfortunately, the reason you cannot reach that better outcome is because the officer suffers from a *commitment problem*. Both sides would be better off if the officer could credibly commit to a quick search; you could then allow a search without worrying that he might renege. However, his words carry no weight; once you consent to a search, the officer can choose any type he wishes. In this game, he prefers to conduct an extensive search to a quick search. And unlike in the game with the boss and his stealing employees, the officer does not earn any benefit from maintaining an honest reputation. As such, if you allow a search, you end up with an extensive search and your worst possible outcome. Therefore, you *must* begin the game by waiting for the K-9.

An important element of a commitment problem is the time inconsistency issues a player faces. Notice what happens when we reverse the order of the moves:

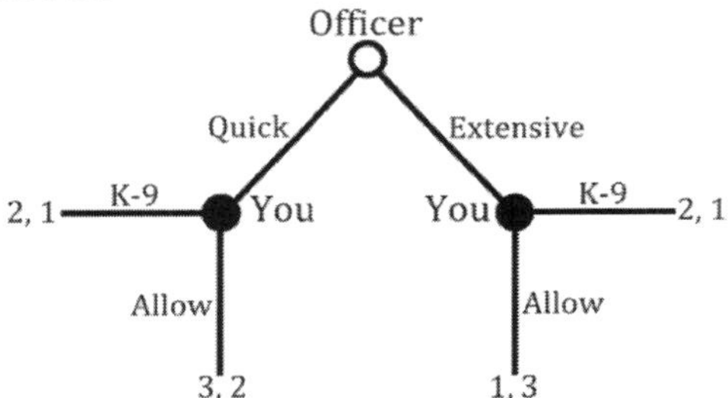

(Note: Normally, the player who moves first has the first payoff and the player who moves second has the second payoff. However, to keep things consistent with the original police search game, your payoffs still come first here.)

Here, the officer begins by choosing a quick or extensive search. This initial choice is binding. You then decide whether he can execute that type of search or wait for the K-9.

Let's quickly solve for this game's SPE, starting with what you should do after the officer selects a quick search:

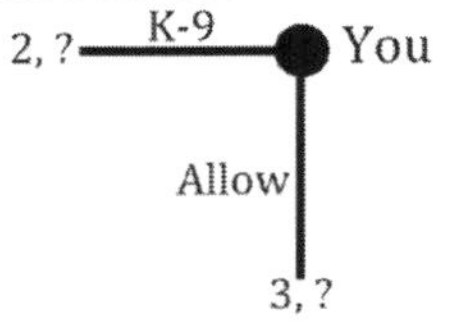

Since 3 beats 2, you allow the officer to conduct his search.

Now suppose the officer chooses an extensive search:

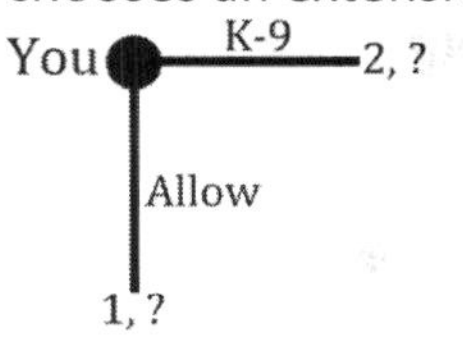

Here, you deny his search and wait for the K-9 unit to arrive, as 2 beats 1.

Knowing these two endgames, we can then solve for the officer's optimal strategy to begin the game:

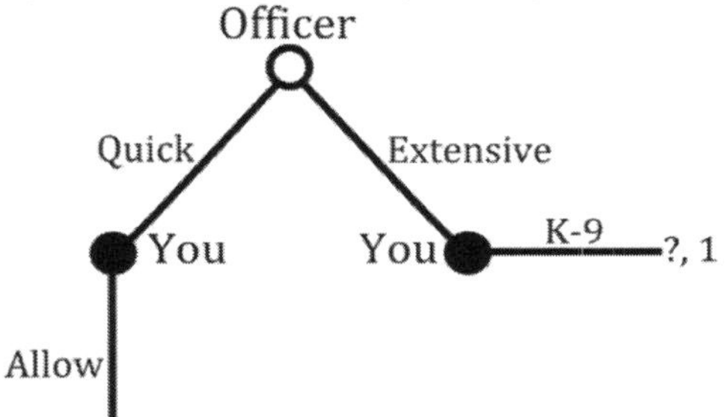

If the officer opts for a quick search, you allow it, and he earns 2. But if he opts for an extensive search, you wait for the K-9, and he earns 1. Since 2 beats 1, the officer selects a quick search.

With the order of moves reversed, the officer and you arrive at the mutually better outcome. But since you are stuck to the original extensive form of the game, the police officer cannot credibly commit to a quick search, and you must wait for the dogs.

The lesson here is that preferences matter more than words. What someone says they will do in the future may be inconsistent with what they would want to do once it is time to follow through. That is not to say words are completely irrelevant. Indeed, the boss's threat to fire his employee worked in that lesson precisely because he made an audible threat. However, the difference between the stealing game and the police game is what happens when the boss and the police officer actually move. When the police officer chooses a search, he *wants* to choose an extensive search because it maximizes his chances of finding illegal materials. He does not care if you think he is a liar. In contrast, the boss *wants* to follow through on his threat, because he cares whether his other employees believe he is fair and honest. He has no time inconsistency problem because he does not want to be seen as a liar.

2.5.1: Civil War

Civil wars rarely end in negotiated settlements; normally, the sides fight each other until one completely militarily defeats the other. Commitment problems explain why.

Suppose some rebels are fighting a successful revolution. The dictator faces a dilemma. He could surrender and save his country from a lot of bloodshed, or he can gamble and hope the war turns in his favor. If successful, he crushes the opposition. If he fails, the rebel leaders take over the country. At that point, or if the dictator concedes at the start, the new rebel government has to decide whether to forgive the dictator or execute him.

Let's look at the extensive form of the interaction. Notice this game introduces two new features: nature (N) and random moves:

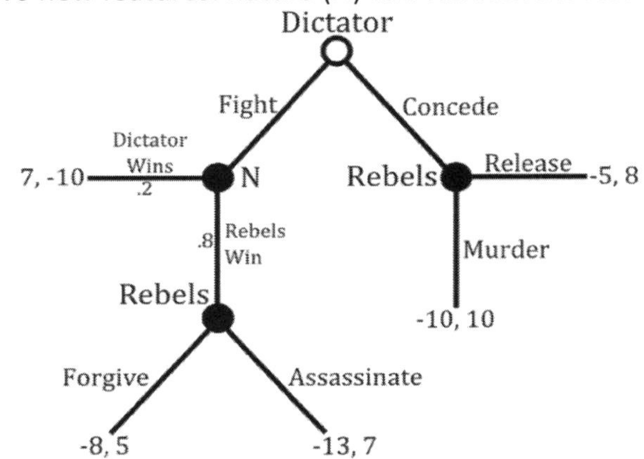

Despite the new complications, we can use backward induction to solve the game. As always, we begin at the end:

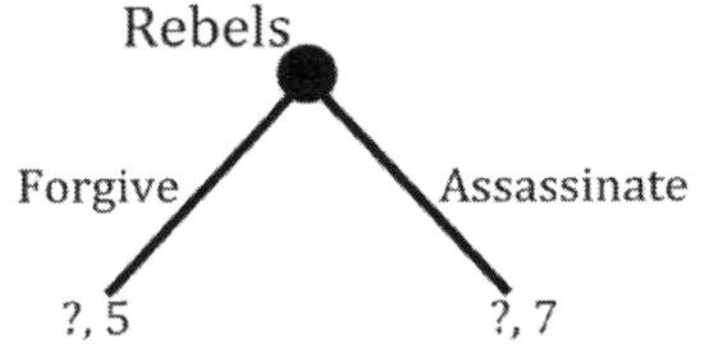

If the rebels win the war, they choose whether to forgive the dictator or assassinate him. Since assassinate is worth 7 and forgive is worth 5, the rebels choose to assassinate the dictator.

Let's look at the next decision node up from there:

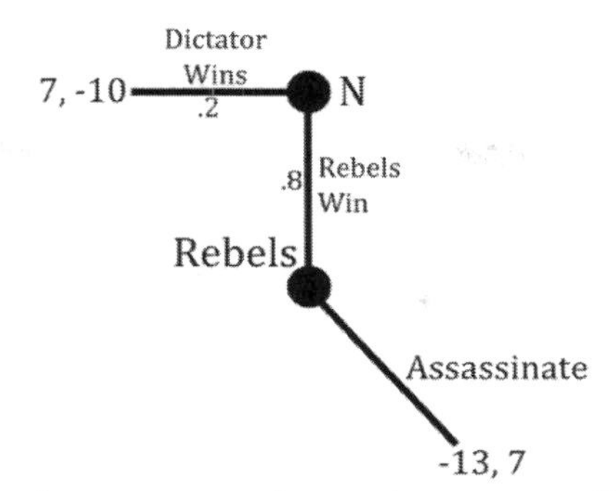

The N represents a move by nature, something we have not encountered before. However, such moves are easy to handle. Nature is not a player; it simply simulates a random movement. Note the .2 below dictator wins and the .8 next the rebels win. These represent the probabilities that nature moves the game into those directions. That is, the dictator wins the war 20% of the time, and the rebels win 80% of the time.

To remove nature from the game tree, we calculate each player's expected utilities for this point of the game, and replace nature's move with those expected utilities. Let's start with the dictator:

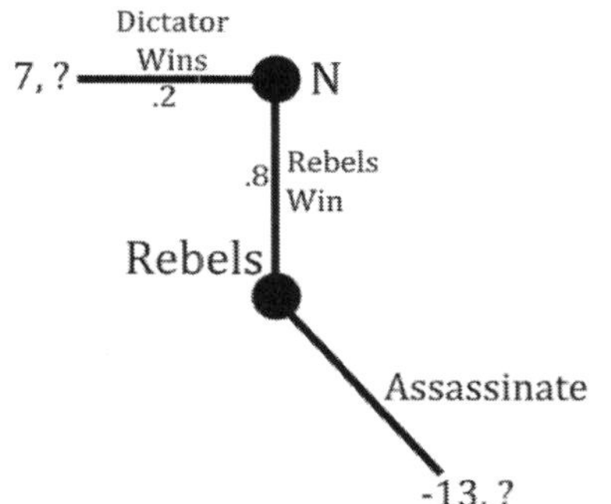

The dictator earns 7 with probability .2 and -13 with probability .8. As an equation:

$EU_{fight} = (.2)(7) + (.8)(-13)$
$EU_{fight} = 1.4 - 10.4$
$EU_{fight} = -9$

Now let's calculate the rebels expected utility if the dictator chooses to fight:

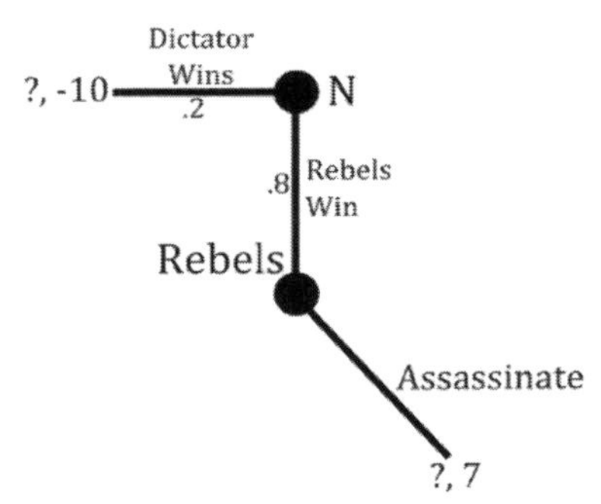

The rebels earn -10 with probability .2 and 7 with probability .8. As an equation:

$EU_{fight} = (.2)(-10) + (.8)(7)$
$EU_{fight} = -2 + 5.6$
$EU_{fight} = 3.6$

Next, we replace nature's move with those payoffs:

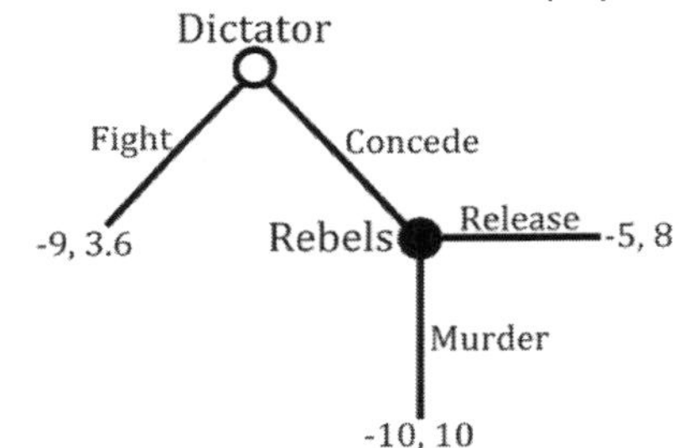

Now the game looks just like any other extensive form we have seen in the past. To finish solving it, we need to find the rebels' optimal move if the dictator concedes.

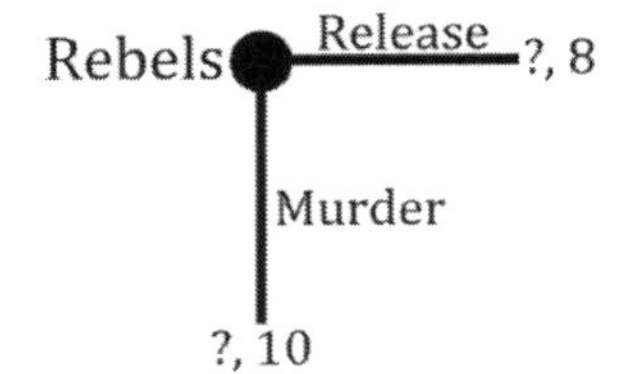

The rebels earn 8 if they release the dictator and 10 if they extract vengeance by murdering him. Since 10 beats 8, they opt for murder.

With all of that backward induction completed, we are down to the dictator's initial decision:

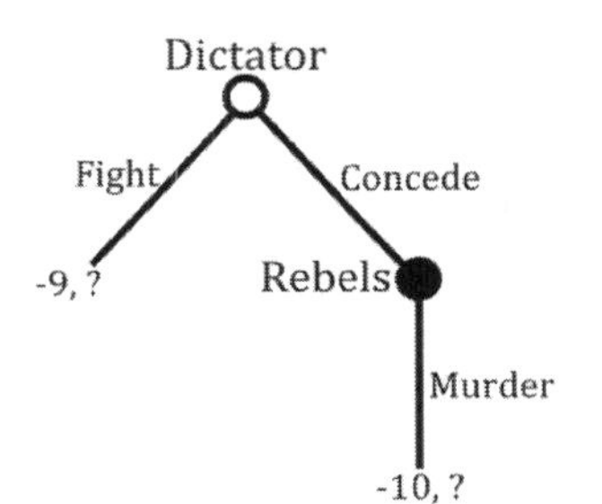

If the dictator fights, he earns -9 on average. If he concedes the war, the rebels murder him and he earns -10. Since -9 beats -10, the dictator gambles on the war. Thus, the SPE is <fight, (assassinate, murder)>, and the sides fight the war to a complete finish.

But suppose the rebels could somehow credibly commit to not killing the dictator if he surrendered:

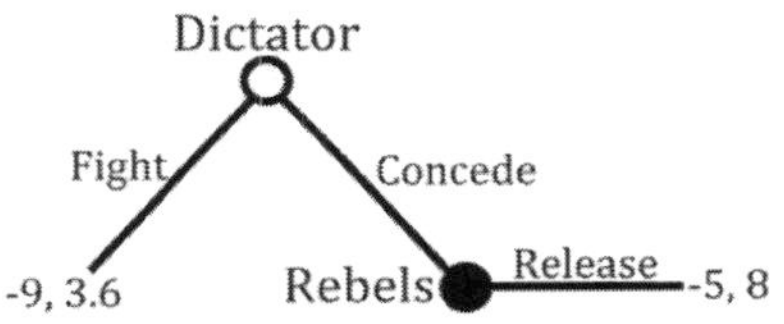

The release outcome is better for both players. The dictator is guaranteed to escape with his life, and his payoff improves from -9 to -5. The rebels minimize casualties and are assured victory, increasing their payoff from 3.6 to 8.

Yet the sides cannot agree to that outcome. Once the dictator concedes the war, the rebels take over power, leaving the dictator's life in the hands of the new government. Knowing that the he will die in that eventuality, the dictator continues the war, even though he wishes he could stop it and escape with his life.

2.5.2: Contracts

Commitment problems also explain the usefulness of contracts. Consider the economic dilemma of a supplier and a manufacturer in the lawless Wild West—a place without a government to legally enforce contracts between two parties. A manufacturer wishes to sell high quality filtered water for $3 per liter. It owns a bottle plant that can produce bottles for $0.50 each, but it does not own a water filtration system. A supplier has the necessary machinery, but the equipment is old and requires $1,000 to repair. Once fixed, the supplier can filter water at a cost of $0.50 per liter.

Negotiations originally stalled when the manufacturer announced that it wished to purchase a total of 1000 liters of water at $1 per liter. The supplier calculated its net profit and declined the offer. If it sold 1000 liters of water to the manufacturer at $1 per liter, they would only take in $1000. However, filtering 1000 liters costs the supplier $500 and the repairs cost $1000. All told, that amounts to $1500 in costs for $1000 in revenue, for a net profit of -$500. The supplier explained this issue to the manufacturer, and the manufacturer responded by upping its offer to $2 per bottle.

Suppose both sides solely want to maximize their profits. Should the supplier accept the $2 offer?

On first pass, the answer appears to be yes. After all, 1000 liters for $2 a liter brings in $2000. After subtracting the $1500 in costs, the supplier stands to make a $500 profit.

Yet the supplier should still refuse the offer. The game tree explains why:

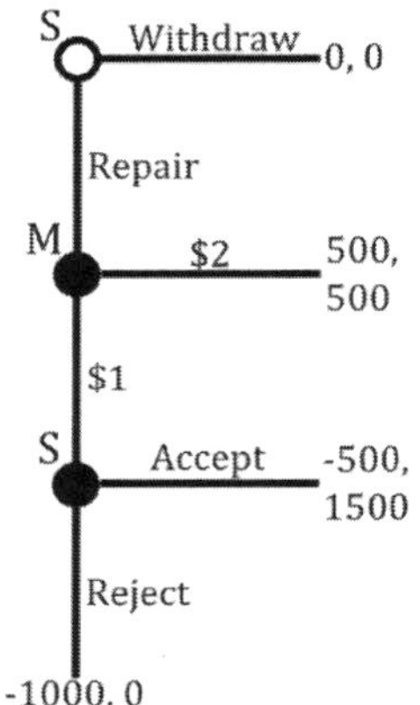

The supplier (S) begins by repairing its machines or withdrawing from the market. If the supplier repairs, the manufacturer (M) decides whether to purchase the liters of water at the agreed upon $2 per liter or renege and only pay $1 per liter. If the manufacturer only pays $1 per liter, the supplier decides whether to accept or reject the revised offer.

Backward induction shows that the manufacturer has incentive to renege on the original agreement. To see this, let's start at the end of the game, after the supplier has made repairs and the manufacturer has offered only $1 per liter:

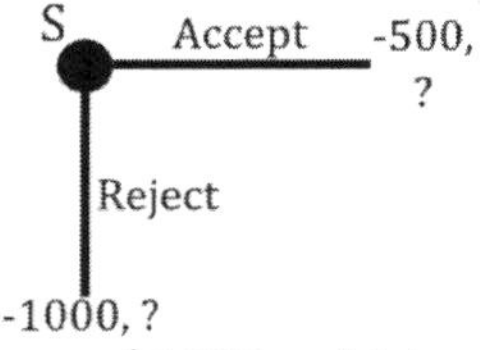

If the supplier quits, it loses $1000, which reflects the cost of repairs it made at the beginning. If the supplier produces, it recoups $500—the manufacturer pays $1000 for water that only costs the supplier $500 to produce. However, the supplier still had to pay $1000 to repair the machines at the beginning, so the supplier loses $500 overall. Since losing $500 is better than losing $1000, the supplier produces in this case.

Let's take that information and move a step backward in the game tree:

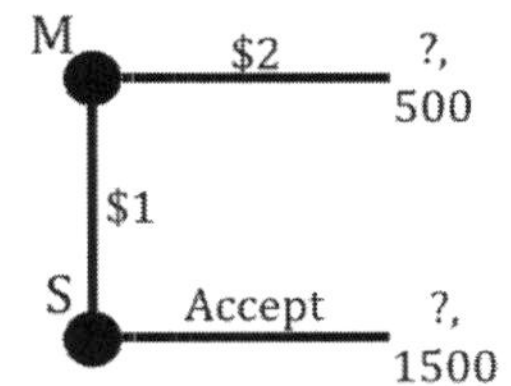

If the manufacturer offers $2 per liter, it makes $500; it sells $3000 of water to consumers, but the bottles cost $500 and the manufacturer has to pay $2000 to the supplier for the water. If the manufacturer offers $1 per liter, the company saves $1000 in payments to the supplier and makes $1500 instead. Since $1500 is more than $500, the manufacturer lowballs the supplier with an offer of $1.

That takes us to the beginning of the game:

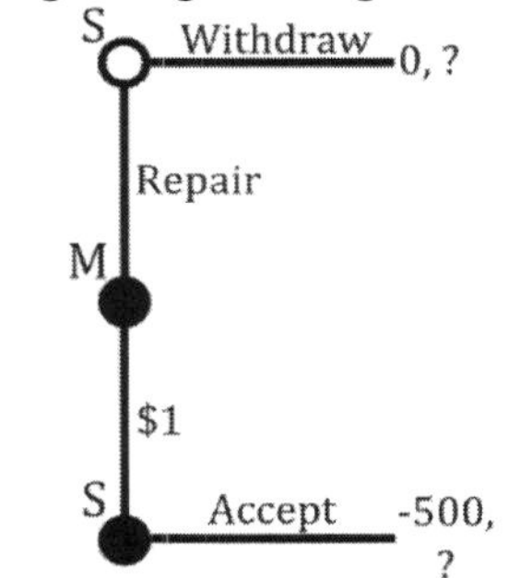

If the supplier withdraws at the beginning of the game, it makes nothing. In contrast, if it repairs its machines, the manufacturer offers only $1 per liter afterward, and the supplier accepts the offer. The supplier loses $500 for this outcome. As such, the supplier withdraws from the market. Thus, the subgame perfect equilibrium is <(withdraw, accept), $1>.

Once more, this is an unfortunate outcome. Compare the equilibrium result to the outcome where the manufacturer sticks to its original offer:

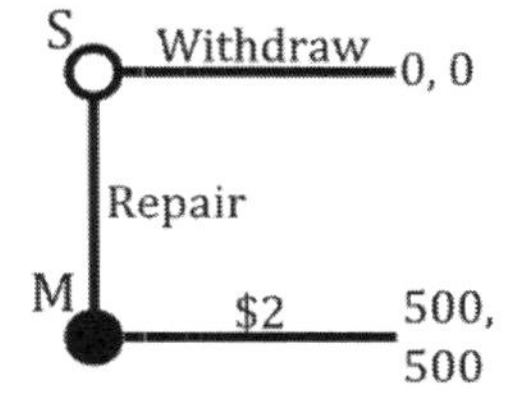

Both sides earn nothing if the supplier withdraws immediately. However, if the manufacturer follows through on the $2 offer, both sides make $500.

The easiest way to resolve this commitment problem is to move out of the Wild West. In a modern day city, the supplier and manufacturer would simply sign a binding contract, specifying that the manufacturer will buy the water at $2 per liter. If the manufacturer reneges, the supplier can sue the manufacturer in court and recoup its lost profit. The Wild West offers no such formal recourse, which allows the manufacturer to exploit the supplier if it repairs its machines.

Barring sudden civilization, a couple creative methods can still solve the problem. The manufacturer could simply buy the supplying company. At that point, the manufacturer does not need to worry about purchasing filtered water, as it would own the means to produce its own. Alternatively, the sides could bring in a third party—perhaps a trustworthy bank—to oversee the transaction and hold funds in escrow. Either of these methods works. The supplier just cannot naively rely on the manufacturer to uphold its end of the deal.

Once again, the lesson here is that preferences and incentives matter more than words. Players must have incentive to follow through on their promises for their words to carry any weight.

Takeaway Points

1) Players encounter a *commitment problem* when equilibrium payoffs are worse for both players than a non-equilibrium outcome.
2) Enforceable contracts can resolve commitment problems, but such contracts are not always available.

Lesson 2.6: Backward Induction without a Game Tree

Up until now, we have looked at some simple extensive form games. In these games, each player had a small number of strategies to select from, which made drawing out the game trees a reasonable exercise. The games in this lesson are not like that. Nevertheless, we can still solve these games through backward induction. The trick is figuring out how to set up the games in such a way that backward induction is effective.

2.6.1: Pirates!

The Dread Pirate Nash captures 10 pieces of gold from the Selten, the Saltwater Scoundrel. He must decide how to divide the coins among the four other members of the crew.

According to pirate tradition, the captain proposes a division of the coins to his crew. If at least half of the crew (captain included) accepts the offer, coins are divided according to the proposal, and bargaining ends. If a majority rejects the proposal, however, the captain must walk the plank. Afterward, the second in command takes over as captain and proposes a new division with the same rules as before. Bargaining continues until all of the pirates are dead or at least half accept an offer.

The Dread Pirate Nash, Pirate 2, Pirate 3, Pirate 4, and Pirate 5 primarily want to survive. Given their survival, they then want to maximize their share of the gold coins. And given a certain allotment of coins, they prefer having that number and having a higher rank in the chain of command than having that same number and a lower rank.

Assume that voting is non-strategic; that is, the pirates always vote according to their preferences. (Although having strategic voting would not change the game's outcome, it does make an already complicated game even more convoluted.) Find the subgame perfect Nash equilibrium.

At this point, we would normally draw the game tree, start at the bottom, and work our way backward until we found every sequentially rational move. However, the actual game tree is incredibly large. To see this, consider the range of offers Pirate 4 can make if Nash, Pirate 2, and Pirate 3 walk the plank. There are 11 possible offers: (10, 0), (9, 1), (8, 2), (7, 3), (6, 4), (5, 5), (4, 6), (3, 7), (2, 8), (1, 9), and (0, 10), where the first number is Pirate 4's share and the second number is Pirate 3's share. Therefore, at this decision node alone, we have eleven different branches coming out of the game tree.

Things only get worse earlier in the game. Pirate 3's possible offers include every way to split 10 items three ways. Drawing such a game tree is simply impractical.

So what now? Rather than drawing an exact game tree, consider the general flow of the game instead:

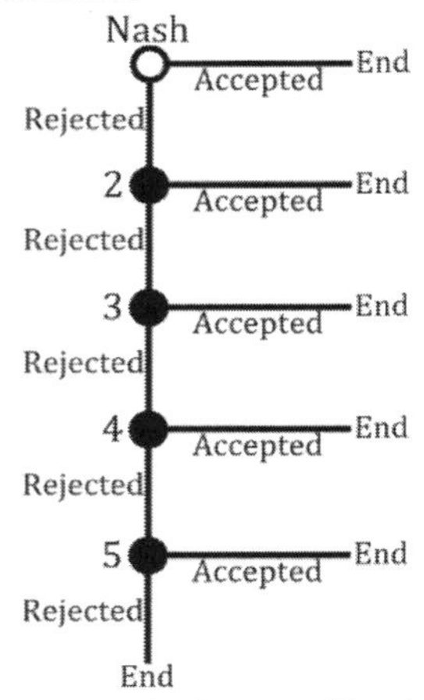

While this tree lacks numerical payoffs, it showcases the order of moves. Since this is a sequential game of complete information, we can still apply backward induction here. The trick is that we must consider all possible offers at every node even though the game tree does not explicitly list them. Once we have Pirate 5's subgame perfect equilibrium offer, we take information and see how it affects Pirate 4's offer. From there, we use that information and apply it to Pirate 3's decision. We repeat this process for Pirate 2 and Nash. At that point, we will have found what occurs in the game's subgame perfect equilibrium.

Let's begin by considering Pirate 5's possible offers. Pirate 5 is the only pirate alive when he makes an offer, so he has to give himself all ten gold coins. Since he primarily does not want to die, he will obviously vote for his own offer, so the game ends with him as captain with 10 gold coins in his pocket. Let's write this down:

Pirates!		Offers Received				
		Nash	2	3	4	5
SPE Proposal	Nash					
	2					
	3					
	4					
	5	Dead	Dead	Dead	Dead	10

Each row represents a subgame perfect proposal for the player in question, while the columns show the offers received for the particular player. We know that if Pirate 5 makes a proposal, he keeps all of the gold coins. The rest of the pirates are dead. Thus, unless a proposal passes before Pirate 5 makes his move, the outcome in row five is the overall outcome of the game.

Now let's take that information and see how it affects Pirate 4's proposal. He knows he dies if the voters reject his offer. Fortunately for him, he only needs at least half of the pirates to vote in favor of the proposal. Since only Pirate 4 and Pirate 5 are alive at this point, Pirate 4's vote is sufficient for the proposal to pass. Consequently, he can assign 10 gold coins to himself, leaving Pirate 5 with nothing. Pirate 5 rejects the offer, but his vote is moot. Pirate 4 cannot possibly improve his outcome, as he is alive, has all of the coins, and is captain. Therefore, he takes everything with his subgame perfect offer.

Pirates!		Offers Received				
		Nash	2	3	4	5
SPE Proposal	Nash					
	2					
	3					
	4	Dead	Dead	Dead	10	0
	5	Dead	Dead	Dead	Dead	10

The game gets complicated with Pirate 3's offer. Unlike Pirate 4's or Pirate 5's proposal, Pirate 3 needs two votes for his proposal to pass. As such, if he has any hope of surviving, he must buy off either Pirate 4 or Pirate 5.

Can he convince Pirate 4 to vote with him? Definitely not. If Pirate 3's proposal does not pass, Pirate 4 knows he survives, becomes captain, and takes all of the coins. This is Pirate 4's best possible outcome; even if Pirate 3 offered Pirate 4 all of the gold coins, Pirate 4 would still reject, knowing that he can obtain the same number of coins by forcing Pirate 3 to walk the plank.

As such, Pirate 3's life is in the hands of Pirate 5. Although we may initially think Pirate 3 has to give most of the coins to Pirate 5 to convince Pirate 5 to vote for Pirate 3, this is unnecessary. Despite Pirate 3's sticky situation, Pirate 5 is in a tough spot as well. If Pirate 3 walks the plank, Pirate 4 gives him no coins at all. Thus, a single gold coin is sufficient to buy Pirate 5's vote; giving Pirate 5 any more coins merely reduces Pirate 3's allotment.

Therefore, Pirate 3 gives Pirate 5 one gold coin and keeps nine coins to himself, he survives, and he finishes as captain. The only way he could improve is if he had all ten coins to himself, but taking a coin away from

Pirate 5 leads to Pirate 3 walking the plank. Consequently, Pirate 3's equilibrium offer is nine for himself, zero for Pirate 4, and one for Pirate 5.

Pirates!		Offers Received				
		Nash	2	3	4	5
SPE Proposal	Nash					
	2					
	3	Dead	Dead	9	0	1
	4	Dead	Dead	Dead	10	0
	5	Dead	Dead	Dead	Dead	10

The chart elucidates the pirates' voting decisions. Look at row three, which represents Pirate 3's equilibrium offer. If the proposal fails, the pirates know they will end up with the outcome in the row directly below. Thus, Pirate 3 decides between nine gold coins or death. Since he prefers life to death primarily, he votes for his own proposal. Pirate 4 earns zero gold coins if Pirate 3's proposal is successful and ten if it fails. He is alive in both cases and prefers more gold coins to fewer, so he votes against the proposal. Finally, Pirate 5 earns one coin for Pirate 3's offer and zero gold coins if Pirate 3 walks the plank. Pirate 5 survives in both cases and prefers more gold coins to fewer, so he accepts the offer. That means two accept and one rejects, so the proposal passes.

Now we move to Pirate 2's offer. Pirate 2 needs at least half of the votes to survive. Since he will vote for his own offer, he needs to buy a single vote from one of the other pirates. Which should he choose? Pirate 3 earns nine gold coins if Pirate 2 walks the plank, so Pirate 2 would have to give Pirate 3 all ten coins to buy that vote. While this ensures Pirate 2's survival, he might want to shop around to see if he can coerce a superior deal.

Let's try Pirate 5. Pirate 5 earns 1 gold coin in equilibrium if Pirate 2 dies, so it only takes two gold coins to buy Pirate 5's vote; one is insufficient, as Pirate 5 would rather have Pirate 2 dead and take the one gold coin that Pirate 3 offers him later. As such, Pirate 2 would much rather buy Pirate 5's vote than Pirate 3's.

However, Pirate 4 offers an even better deal. In this round of voting, he is the most vulnerable—if Pirate 2 walks the plank, Pirate 3 offers him zero coins in the next stage. As a result, Pirate 4 would vote to keep Pirate 2 alive as long as his share is at least one gold coin. Pirate 2 is happy to oblige, giving Pirate 4 exactly one coin; any more is unnecessary to buy Pirate 4's vote and only decreases Pirate 2's payoff. So Pirate 2 assigns himself nine coins and Pirate 4 one coin; Pirate 2 and Pirate 4 vote for the proposal, and the game ends.

Pirates!	Offers Received				
SPE Proposal	Nash	2	3	4	5
Nash					
2	Dead	9	0	1	0
3	Dead	Dead	9	0	1
4	Dead	Dead	Dead	10	0
5	Dead	Dead	Dead	Dead	10

Finally, we must find Nash's optimal offer. To survive, he must obtain two votes from the other four pirates. Looking at the chart, Nash observes that Pirate 3 and Pirate 5 receive zero gold coins if he walks the plank. Therefore, he can give one coin to each of those Pirates, and both will approve the offer. Nash survives and keeps the remaining eight gold coins in this case.

Nash cannot do better. If he tries to buy off Pirate 2, he must offer all ten coins. That leaves no coins left to buy the third vote. Nash could buy Pirate 4's vote with two coins and give one coin to Pirate 3 or Pirate 5, but he only receives seven coins by including Pirate 4 in the bargain. Thus, giving one coin to Pirate 3 and another coin to Pirate 5 is Nash's best option.

Pirates!	Offers Received				
SPE Proposal	Nash	2	3	4	5
Nash	8	0	1	0	1
2	Dead	9	0	1	0
3	Dead	Dead	9	0	1
4	Dead	Dead	Dead	10	0
5	Dead	Dead	Dead	Dead	10

This final table gives all the details of the subgame perfect equilibrium. However, in the equilibrium, only Nash makes an offer. Nash, Pirate 3, and Pirate 5 approve it, and the game ends. By thinking things through, Nash escaped the plank and ended up doing quite well for himself.

2.6.2: Nim

21 poker chips sit between two players at a table. Player 1 begins by taking one or two chips from the stack. Then player 2 selects one or two. The players continue to alternate in this fashion. The player who takes the last poker chip from the stack is the winner. Who will win this game?

Once again, drawing a game tree is nearly impossible; each player has two choices at every decision node, which leads to a ton of possible outcomes. Rather than analyze each of these outcomes individually, we can benefit by grouping similar decisions together.

For example, consider the trivial case where a player moves with only one chip left in the pile. That player must take the remaining chip, and he therefore wins the game. This covers *all* cases where there is only one chip remaining, regardless of the history of the game or whose move it is.

Now consider a similarly trivial case where a player moves with two chips left in the pile. This subgame is relatively tractable, so we can draw the following game tree to represent the situation:

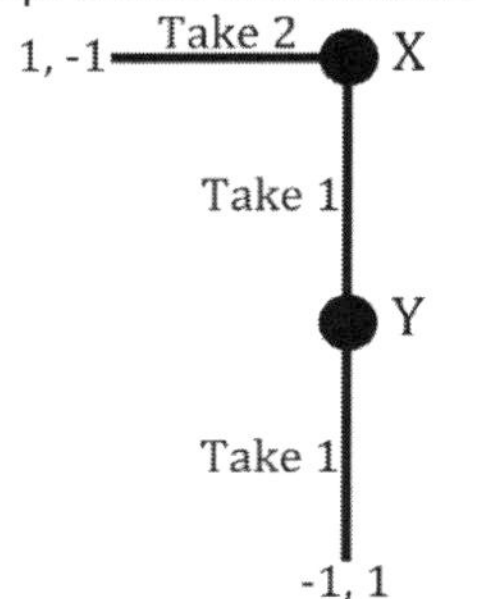

Note that we have generalized the players in this abstract game tree, which rewards 1 for a win and -1 for a loss. It does not matter whether player 1 or player 2 is X. (For the sake of clarity, however, player X's payoffs will always come first in this section.) Player X simply refers to the person who moves whenever only two chips are left. If player X takes two, he wins and earns 1. If he takes one, player Y has to take one as well. The end result is that player X loses and earns -1. Since 1 is greater than -1, player X ought to take two.

Let's move back a step. Suppose player Y has three chips left in the stack. What should he do?

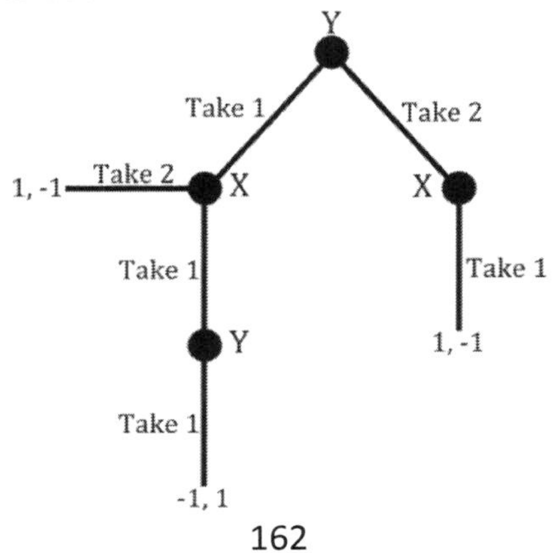

If player Y takes two, player X takes one and wins. Alternatively, if player Y takes one, player X takes two and wins. (We know from the last step that player X would not take one, as player Y then takes one, resulting in player X losing.) Thus, player Y is in a no-win situation. Regardless of whether he takes one or two, player X can always counter with a winning strategy. In turn, player Y ultimately earns -1 regardless of his decision here.

Now imagine there are four items left in the stack. Can this person win? Yes. By taking one, that player leaves three in the stack for his opponent, and we know that a person loses if three items are left in the pile. Alternatively, that player could take two and leave two remaining, but that means his opponent can take two and win the game. Drawing out the game tree for this would be extremely cumbersome; merely working through the logic of backward induction saves us time here.

What if there are five? The player could take two and leave his opponent with three, thus guaranteeing the win. Taking one is a losing strategy, as the opponent could respond by taking one as well, thus leaving the original player with only three left in the pile, which is a losing position. So with five left, the rational player takes two.

And six? The player is in trouble. If he takes one, his opponent can respond by taking two, leaving him with three left in the pile, so he loses. Likewise, if he takes two, his opponent can take one, leaving him with three left once again. As such, regardless of his strategy here, the player loses.

How about seven? The player can take one and leave his opponent with six, guaranteeing a win. On the other hand, he could grab two, but then his opponent could grab two as well. That leaves him with three left in the pile, so he loses. Consequently, he can only take one.

If there are eight in the pile, however, he must take two. That leaves six left for his opponent, guaranteeing him victory. If he takes one, on the other hand, his opponent can also take one, leaving six in the pile and a loss for the original player.

Nine? This is the same position as six. If the player takes one, his opponent takes two, and he is stuck with six in the pile and a loss. Taking two does not improve his position; the opponent takes one instead, he is left with six, and he still loses.

By now, the pattern should be obvious. If a player moves when there is a multiple of three left, then he can select whatever he wants, and his opponent follows it up with the opposite action. So if the player takes one, the other play takes two; and if the first player takes two, the second player takes one. In either case, three poker chips leave the pile, leaving a multiple of three items still left. Whoever makes the second play is guaranteed to take the last chip from the stack.

In contrast, when the number of chips left is not a multiple of three, the player can reduce the pile to a multiple of three, which ultimately leads to victory.

So who wins this game? Since there are 21 poker chips in the original stack, the second player wins. Her path to victory is simple: whenever player 1 moves, she makes the exact opposite action. After two moves, 18 chips will be left. After four, it will be down to 15, and so forth. Eventually player 2 takes the final one or two chips left in the stack and wins.

Conclusion

In the pirate game and nim, the difficulty in finding the solution not using backward induction but rather setting up the problem to apply backward induction.

However, some situations do not have immediately obvious endgames. And without an endgame, backward induction is useless. Consequently, when you see a problem of this sort, think about how the interaction could possibly end and whether there is a general pattern to these endgames. Work backward from those points, even if you cannot draw an adequate game tree to operate from. Eventually, you might arrive at a solution.

Lesson 2.7: Problems with Backward Induction

At face value, backward induction and subgame perfection are intuitive solution concepts. However, in extreme cases, backward induction can lead to some strange—and possibly implausible—outcomes. Moreover, rational players may have incentive to deliberately act *irrational* so they can increase their payoffs. This section covers three of those gray areas, reminding us how fragile our assumptions can be.

2.7.1: Mistake-Free Games

To start, actors in our models *never* make mistakes. This is reasonable to assume when there are only a couple of players and two or three moves, but things quickly get out of hand as we increase the complexity of the game.

Consider this situation. One at a time, each of 100 individuals chooses to take or to contribute. If all 100 players in a row contribute, everyone earns 100. However, if any player chooses take at any point in the process, the game immediately ends, the player who chose take earns 1, and every other player earns 0.

We can use backward induction to solve the game. Suppose the first 99 individuals contribute. Then the final decision node looks like this, with only individual 100's payoffs shown:

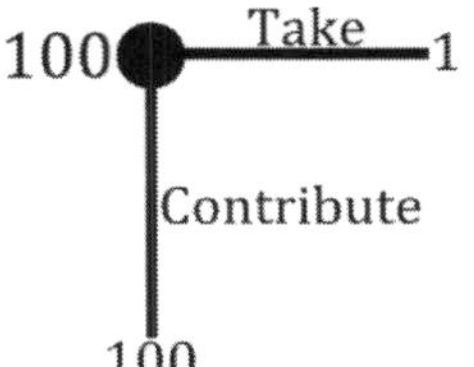

If he contributes, he earns 100. If he takes, he earns 1. Since 100 is greater than 1, he contributes.

Let's roll back one step and find individual 99's optimal play. Here is his choice, this time with only individual 99's payoffs showing:

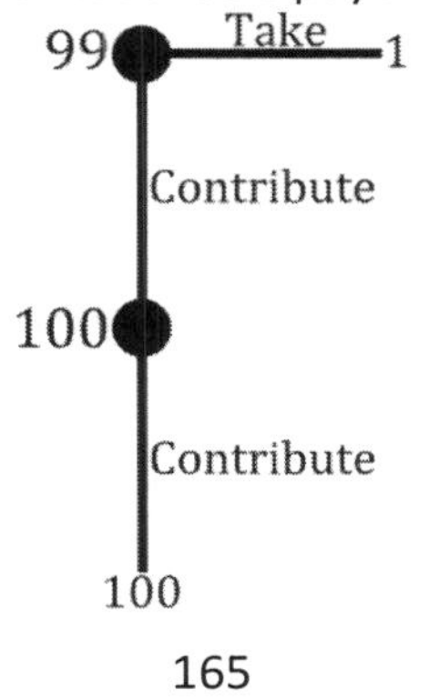

If he takes, he earns 1. If he contributes, then individual 100 contributes, and he earns 100. Thus, he contributes. We could go through backward induction for the remaining 98 players, but we would see the same thing: the players contribute all the way down the line and ultimately all earn 100.

Yet something is unsatisfying about this SPE. While player 100 can contribute in comfort knowing that he will reap the collective benefit, each player moving backward relies more and more heavily on the other players' optimal play to achieve the large payoff.

In that regard, player 1 is in the worst position: he needs the 99 players after him to play optimally to justify contributing. Suppose he believes every player after him is 99% likely to contribute. Then the probability all of them contribute is $.99^{99}$, or slightly less than 37%. Since complete contribution is worth 100 and taking is only worth 1, contributing is still a wise play.

But suppose player 1 held his other players in lower regard. Instead of estimating all of the players to contribute 99% of the time, this time he believes they are 95% likely to contribute. Although this is still an optimistic outlook, the large number of players dooms the outcome. The probability that all remaining players contribute equals $.95^{99}$, which is slightly greater than 0.006%. In that case, player 1 ought to take the 1 upfront and end the game.

Further problems exist. Suppose player 1 estimates that players contribute 99% of the time. However, he believes that player 2 has a grim outlook regarding players 3 through 100, estimating each of them is only 95% likely to contribute. Thus, in player 2's eyes, there is a $.95^{98}$, or slightly greater than 0.0065% chance, that contribution will ultimately pay off. Consequently, player 2 takes her payoff of 1 and ends the game. Player 1 earns 0 for this outcome.

But if player 2 takes, player 1 should preempt player 2 by taking at the beginning of the game; he earns 1 if he takes immediately versus 0 if he waits for player 2 to take. Therefore, player 1 can only rationally contribute if he is extremely optimistic of everyone else's ability to play optimally *and* believes everyone else is optimistic as well. What seemed like a simple choice for player 1 has evolved into a deeply complicated affair!

Unfortunately, backward induction and subgame perfection lead us nowhere on these issues. While this is not an indictment on backward induction or SPE, the game does remind us of their assumptions. We *assume* players are rational. We *assume* they know what is best for them. We *assume* they do not make mistakes. Although these assumptions are not heroic when the game is small and involves only a few players, it becomes further and further dicey with more and more additions to the game. And if we want to explicitly model mistakes under our rationalist framework, then we must put moves from nature into the game that possibly (and randomly) end the game at each step.

2.7.2: Complete Information: The Chain Store Paradox

Backward induction has other issues. This time, suppose a chain store has locations in five different cities and faces a challenge from five individual local companies. All of the potential challengers only want to enter the market if the chain store does not engage in a price war. The chain store finds it more profitable to accede to each of the challenges individually than to start a price war to drive the competitor out of business. However, the chain store would rather start a price war with one or two of the competitors if it deters the other three from entering the market.

The game begins with a challenger in town 1 choosing to enter the market or sit out. If it enters, the chain store chooses whether to engage in a price war or not. Afterward, the challenger in town 2 sees what occurred in town 1 and then decides whether to enter. Once again, if the challenger in town 2 enters, the chain store decides whether to engage in a price war. The game repeats sequentially three more times in the three remaining towns.

The game tree is far too large to fit in these small pages, so we will have to work out the logic of the game without it. Regardless of the previous moves, consider the chain store's final action if the challenger in town 5 enters the market. The introduction of the game mentioned that the chain store finds it more profitable to accede to each challenger individually. So the chain store accedes in this case. Moving back a step, the challenger in town 5 only wants to enter if the chain store accedes. Since it knows the chain store will accede, the challenger enters.

Now consider how the chain store should deal with the challenger in town 4 entering the market, regardless of the history of the game before that. Again, since the chain store finds it more profitable to accede to the challenger, it does not engage in a price war. Therefore, the challenger in town 4 enters.

This logic repeats all the way up the game tree. Ultimately, the subgame perfect equilibrium is awful for the chain store: every challenger enters and the chain store never challenges. Essentially, the challengers completely push over the chain store.

But suppose you are the challenger in town 2, and you see that the chain store has started a price war against the challenger in town 1. Now what should you do? The subgame perfect equilibrium says you should challenge. However, subgame perfection is based on rational play—or, at least how we define "rational" play. When the chain store engaged in a price war in town 1, it demonstrated irrational behavior. Consequently, it is no longer clear whether you should expect a rational response when the chain store decides whether to start a price war with you. If a price war means that you go into deep debt, it may not be worth the risk for you to challenge in the first place.

Now suppose you are the chain store, you are perfectly rational, and you cannot credibly threaten a price war with any potential challengers. Your rival in town 1 challenges you. How do you respond? Your perfect rationality tells you not to start a price war. But imagine you did. As you just saw, your irrational price war against the challenger in town 1 might deter the challenger in town 2 from entering the market. The same could be true for the challengers in towns 3, 4, and 5. That being the case, you ought to start a price war against the first competitor—even if doing so is completely irrational by our standards. In turn, the challenger in town 1 might not want to engage, out of fear that you plan to deliberately act irrational.

Thus, the "chain store paradox" is that backward induction tells us that all the competitors will enter the market, yet we have a perfectly good reason to believe that no competitor will.

Where did backward induction fail us here? Again, the combination of lots of actors, lots of moves, and lots of assumptions sets us up for trouble. Of particular concern here, however, is the concept of complete information. That is, in the model, we assume the chain store finds it unprofitable to start a price war against any of the possible challengers and that the possible challengers know this. Realistically, though, challengers might be unsure whether the chain store maximizes its profits in each individual town by starting a price war or by acceding to the challengers.

Unfortunately, incomplete information game theory is well beyond the scope of this chapter. Nevertheless, we do not need a model to understand that a weak chain store has incentive to pretend that it is a strong chain store by engaging in a price war with any competitor that challenges it. Consequently, one possible resolution to the paradox is that we tried to fit a game of incomplete information into a game of complete information. And if we do not put the correct assumptions into the model, we should not expect to see the correct expectations coming out of it.

However, something is still unsettling about the backward induction prediction even if we stick to the complete information story exclusively. Moreover, the chain store paradox is not the only game with this problem. Here, the chain store deliberately acted *viciously* irrational to improve its payoff. In the next section, we explore a game where players can deliberately act *benevolently* irrational and still reap the benefits.

2.7.3: Feigning Irrationality: The Centipede Game

This is a two player game. Player 1 begins by taking $2 or adding $2 to the pot. If he takes the money, the game ends. If he adds the $2 to the pot, the second player can take $2 and split the $2 in the pot, or she can add the $2 to the pot. If she takes the money, the game ends and player 2 earns $3 and player 1 earns $1. Otherwise, player 1 chooses to take $2 and split the $4 in the pot or add that $2 to the pot. If he takes the $2, the game ends, and he earns $4 while player 2 earns $2. Otherwise, player 2 faces the same decision as before but with a larger pot. This process repeats for 100 total rounds, hence the "centipede" name.

The first few rounds look like this:

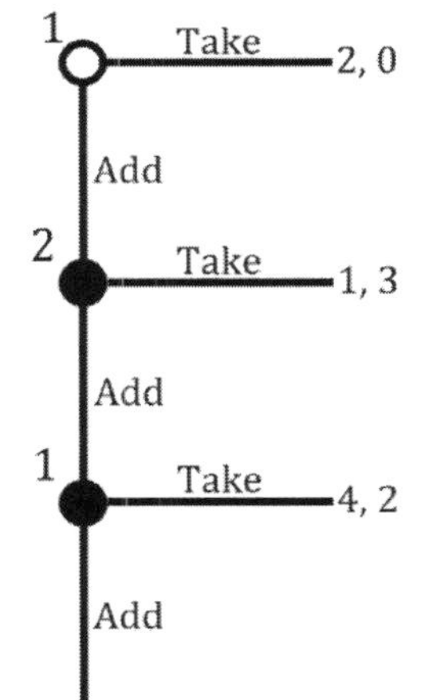

If the players always add the money to the pot, then the game continues until $200 is available. In the final move, player 2 decides whether to take $101 and give $99 to player 1 or split the money so that both players take $100.

The last few rounds look like this:

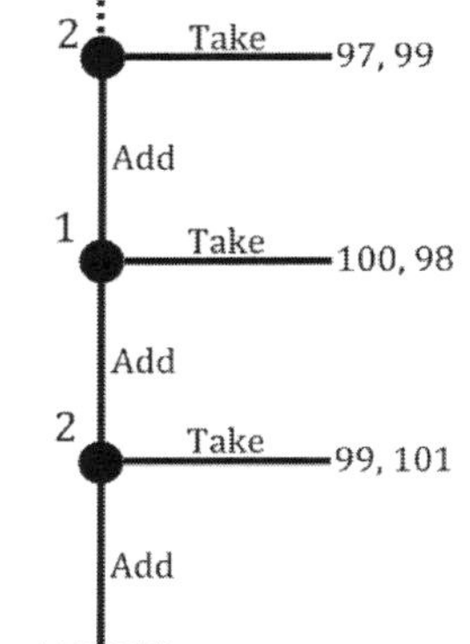

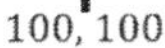

Suppose the players only want to maximize their monetary payoff. Then we can use the dollar figures as expected utilities and solve the game with backward induction. Let's start at player 2's final decision:

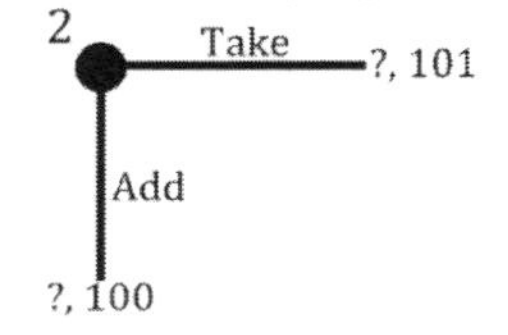

Since 101 beats 100, player 2 takes if the game reaches the final decision node.

Now let's backtrack to the second to last choice:

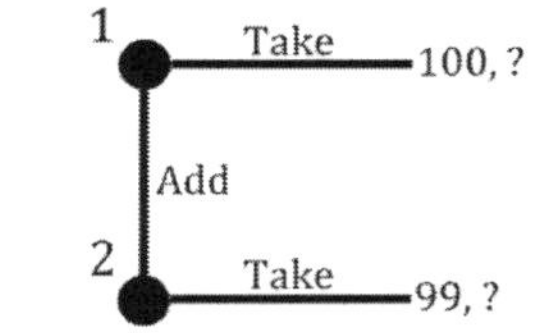

If player 1 adds, player 2 takes, and player 1 earns 99. If he takes, he earns 100. Therefore, he takes.

Let's go back another step:

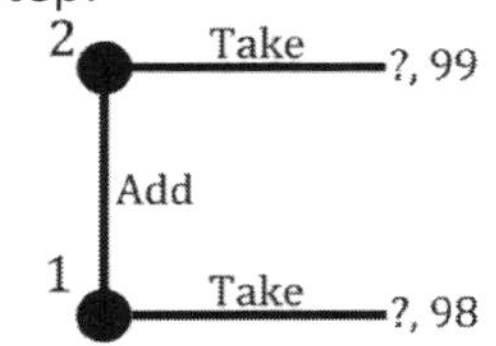

This is a similar story. If player 2 adds, player 1 takes, and player 2 earns 98. If she takes, she earns 99. Therefore, she takes.

The same logic repeats over and over again. Eventually, we reach the first decision node:

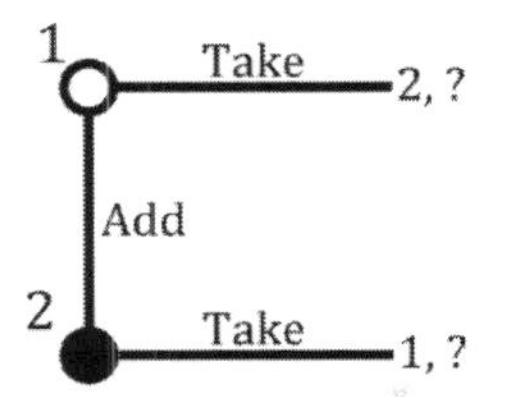

If player 1 begins by adding, player 2 ends the game at the next decision node, and player 1 earns 1. Alternatively, he could take immediately and earn 2. Since 2 beats 1, the game begins and ends with player 1 taking. Thus, the SPE is for each player to take at every decision node. The subgame perfect outcome is unfortunate; many outcomes have better payoffs for both players than what occurs in equilibrium.

The centipede game is a frequent subject of laboratory experiments. Although the game ends immediately in the SPE, in practice players generally play for many rounds before someone finally takes the extra two dollars. As with the chain store paradox, game theorists have a variety of explanations for the discrepancy between subgame perfect play and play in practice.

First, players may be irrational. They simply may be unable to work through the logic to understand that they ought to take immediately. In turn, they repeatedly add until they are close enough to understand the dilemma, which eventually causes the game to end.

Second, note that just a single irrational player can throw a wrench into the system. Suppose you are a rational player 2, and player 1 begins the game by adding. You realize he is not rational and wonder what would happen if you added as well. Given that he has already contributed to the pot, it stands to reason that he will do so again. Moreover, if you chain together a few rounds of cooperation, your ultimate payoff will be substantially larger than the 3 you earn if you end it at your first available opportunity. As such, you may be inclined to add over your first dozen or so decision nodes. Thus, irrational play sparks further irrational play.

Third, this logic in turn destroys the backward induction solution when there are two rational players. Suppose you are a rational player 1 and you ignore the SPE by beginning with add. Now the rational player 2 has no idea what is going on. She may figure you are irrational and so she should continue as in the previous case. Alternatively, she may think you are rational but deliberately acting irrational in hopes that she will play irrationally as well, thus improving both of your payoffs. This time, irrational play sparked further irrational play, yet no one was irrational! What it means to be "rational" here is not crystal clear, so perhaps our very definition of rationality is what fails us.

On the other hand, the centipede game's experimental evidence may rest on a faulty assumption. We used the dollar figures as expected utilities, but it is reasonable to believe that people also value altruism. That being the case, player 1 might find choosing take in the first round to be unacceptably mean. If so, the expected utilities we were using for that game were incorrect, and we should not expect false assumptions to produce accurate predictions.

Takeaway Points

1) As with everything else in game theory, backward induction is only as good as its assumptions.
2) A game in which players do not make mistakes is fundamentally different from a game where players do make mistakes. It should not be surprising that the results are different when we change that assumption.
3) Players sometimes have incentive to feign irrationality, creating opportunities for both players to benefit.

Lesson 2.8: Forward Induction

When we solve games with backward induction, players believe all future play will be rational, and they condition their present behavior on what will occur in the future. Forward induction adds an extra layer of complexity. Here, the players believe that all *prior* play was rational as well, and they condition their present play based off what they can infer about past play.

Although forward induction may seem like a straightforward assumption, it quickly leads to some involved inferences. We will start with a simple example and work our way up in complexity.

<u>2.8.1: Pub Hunt</u>

Let's begin with a modified form of the stag hunt. Player 1 chooses whether to hunt for a stag, hunt for a hare, or go to the pub. If he goes to the pub, player 2 will see him and automatically join him there. Both earn 2.5 for this outcome. If player 1 hunts, player 2 chooses a target as well without seeing what player 1 is aiming for. Their payoffs here are the same as the original stag hunt.

Here is the game tree:

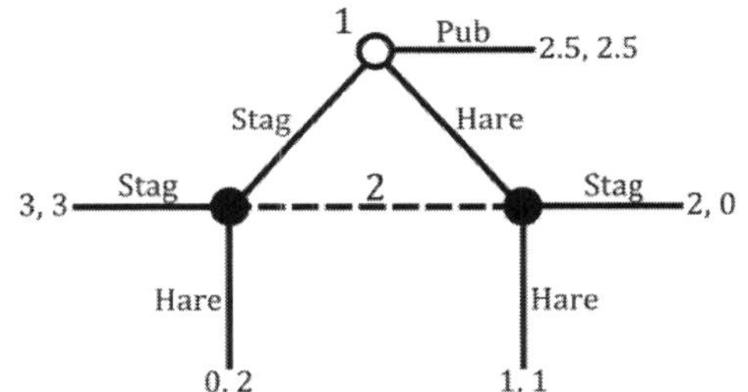

If we ignore player 1's pub move, the remaining game is an ordinary stag hunt:

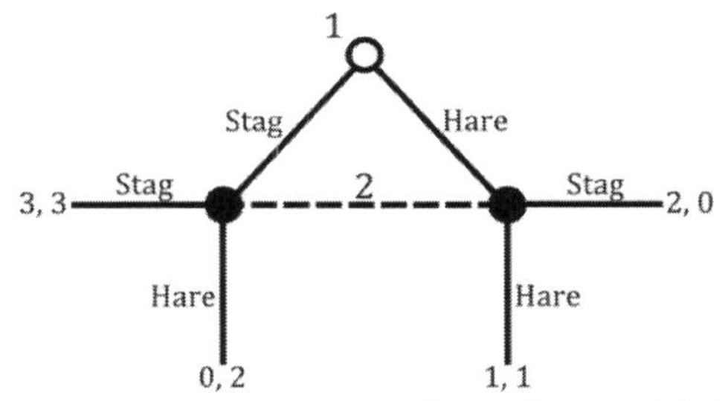

We know a simultaneous move stag hunt has multiple Nash equilibria. This presents a new problem. We have seen simultaneous move games with multiple Nash equilibria before but never when additional moves were around. Here, player 1's pub strategy throws a wrench in the system;

we cannot be sure whether the Nash equilibria of the stag hunt make sense in the presence of the pub strategy.

Let's think our way through this one. Suppose player 1 chose hare. Depending on which strategy player 2 selects, player 1 could end up in either of these two outcomes:

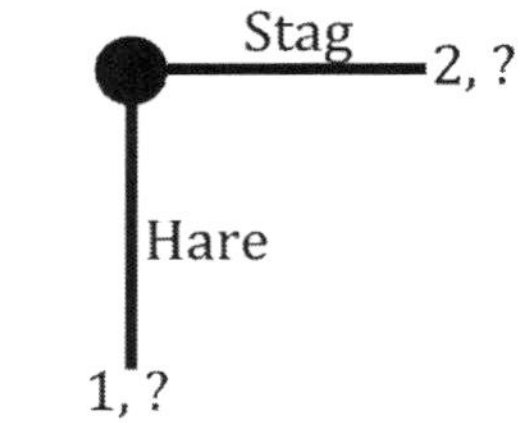

In player 1's best case scenario, player 2 hunts a stag, leaving player 1 with a payoff of 2. In the worst case scenario, she chases a hare as well, and he earns 1.

Compare those outcomes to going to the pub instead:

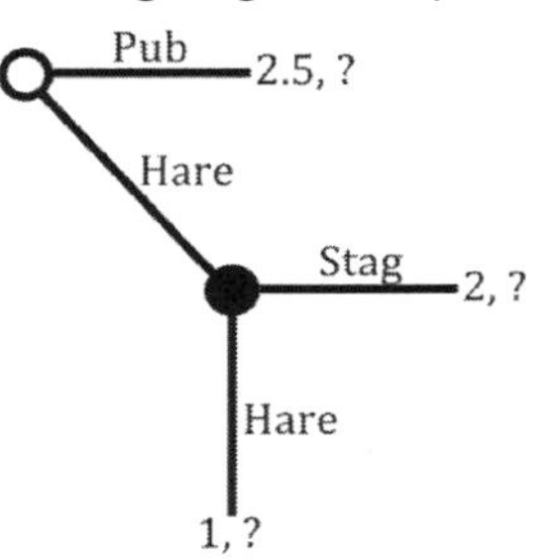

Going to the pub locks in a payoff of 2.5 for player 1. In contrast, the most he earns if he plays hare is 2. Thus, it is irrational for player 1 to play hare.

Now consider player 2's position. Suppose she has an opportunity to move. The game looks like this from her perspective:

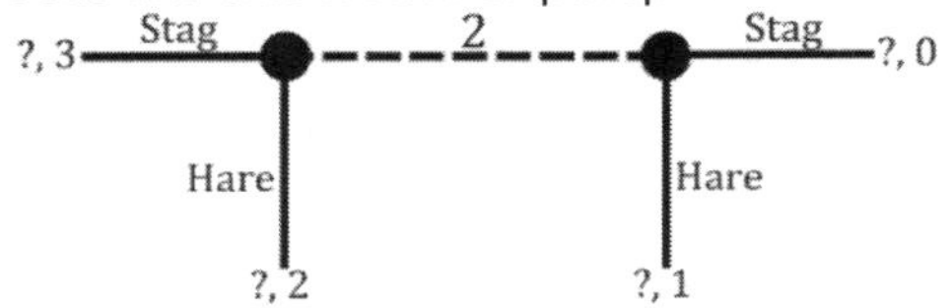

Player 2 does not know whether player 1 chose stag or hare; all that she sees is that he did not go to the pub. If he chose stag, she should choose stag as well; but if he aimed for hares, she should follow suit. All around, player 2 is in a dilemma.

Or is she? According to forward induction, all past play was rational. We saw that player 1 would never rationally play hare. Thus, although player 2 cannot directly see player 1's move, she can infer he did not play hare and therefore played stag. In turn, she can narrow down the path of play to this:

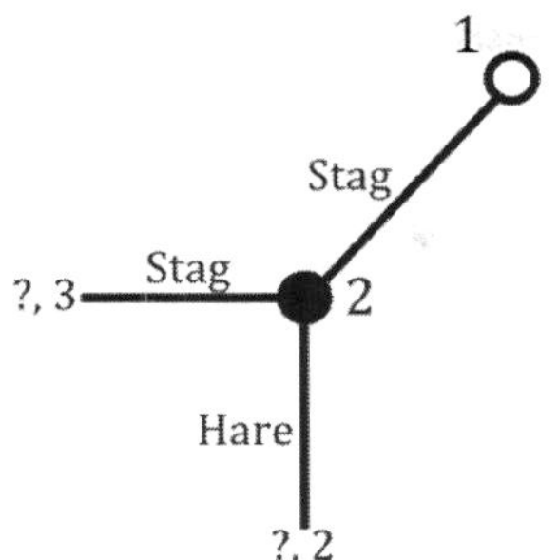

Since 3 beats 2, she should hunt a stag.

Now let's work through player 1's move. He can reason through the above and conclude that player 2 will play stag. Thus, his decision looks like this:

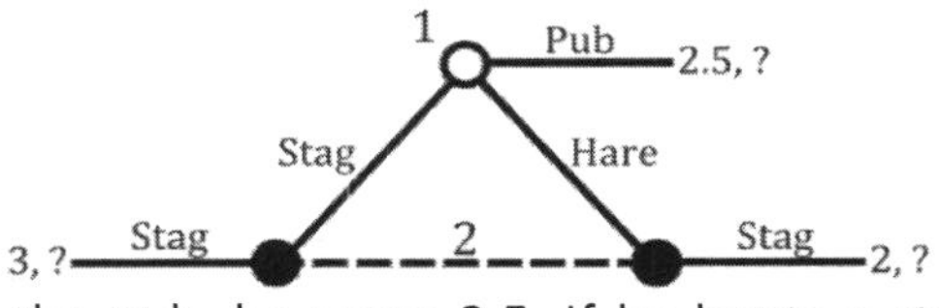

If he goes to the pub, he earns 2.5. If he hunts a stag, player 2 joins him, and he earns 3. Finally, if he hunts a hare, player 2 still aims for the stag, and so he earns 2. Since 3 beats both 2.5 and 2, he hunts a stag. Conveniently enough, forward induction has left us with <stag, stag> as the unique equilibrium.

After boiling the process down, we witnessed a complicated application of strictly dominated strategies. Let's look at the matrix of the pub hunt:

	Stag	Hare
Stag	3, 3	0, 2
Hare	2, 0	1, 1
Pub	2.5, 2.5	2.5, 2.5

Specifically, compare player 1's pub strategy against his hare strategy:

	Stag	Hare
Hare	2, ?	1, ?
Pub	2.5, ?	2.5, ?

2.5 beats both 2 and 1. Therefore, regardless of player 2's strategy, player 1 earns more at the pub than he does hunting a hare. This is our definition of strict dominance; accordingly, we ought to remove player 1's hare strategy from the matrix:

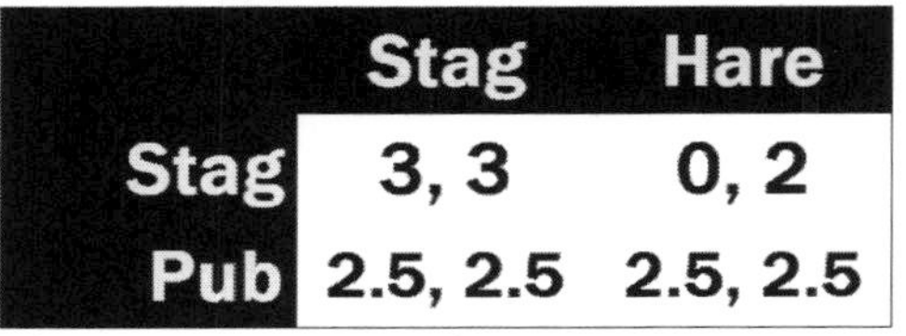

	Stag	Hare
Stag	3, 3	0, 2
Pub	2.5, 2.5	2.5, 2.5

If we mark the best responses to the remaining game, we find two pure strategy Nash equilibria:

	Stag	Hare
Stag	3*, 3*	0, 2
Pub	2.5, 2.5*	2.5*, 2.5*

So how did we only end up with one solution via forward induction? We could think of the matrix sequentially. Player 1 moves first. He can either play pub or stag. Consider player 2's best response to pub:

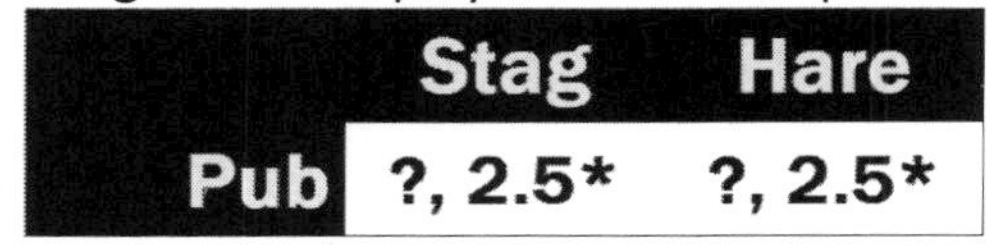

	Stag	Hare
Pub	?, 2.5*	?, 2.5*

Player 2 earns 2.5 regardless of her choice, so she is free to choose either strategy. (In practice, the game ends when player 1 picks pub, so player 2 never actually moves. This is why player 2 is indifferent between her strategies and both players earn the same payoffs.)

Now consider player 2's best response to stag:

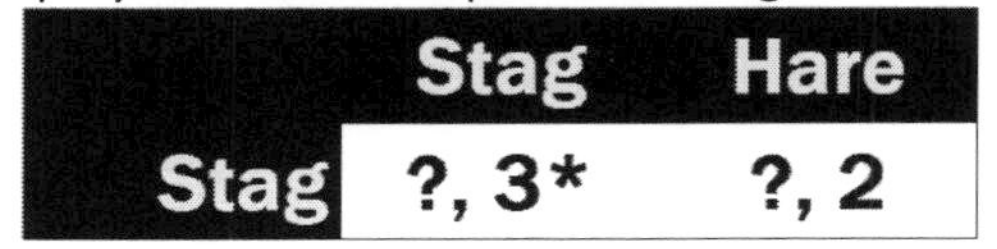

	Stag	Hare
Stag	?, 3*	?, 2

Here, she chooses stag.

Forward induction effectively allows player 1 the choice of which universe he wants to live in: the universe in which player 2 responds to him going to the pub or the universe in which player 2 responds to him hunting a stag. He earns 2.5 in the first universe and 3 in the second universe, so naturally he chooses to live in the second universe. Thus, <stag, stag> is the unique equilibrium.

Although logically demanding, forward induction leads to a plausible result here. After all, in the original stag hunt, the players merely wanted to coordinate on the stag. Introducing the pub allows them to do this, even though the players never meet there.

Unfortunately, the pub hunt is also the simplest application of forward induction. The examples grow increasingly bizarre from here.

2.8.2: Defenestrated Chicken

In Lesson 1.6, we solved the game of chicken, in which two teenagers drove straight at each other, waiting for the other one to swerve. We now consider a richer form of the game. Player 1 has a trick steering wheel that can be removed at will. The game begins with player 1 choosing whether to toss his steering wheel out the window—that is, defenestrate it—or not. If player 1 defenestrates his steering wheel, he is physically incapable of swerving. Player 2 observes the toss and recognizes that player 1 has locked into continuing, and she chooses whether to continue or swerve. If he keeps his steering wheel in, he chooses whether to continue straight or swerve. Player 2 must also choose between continuing straight and swerving, though she cannot see player 1's strategy.

Here is the game tree:

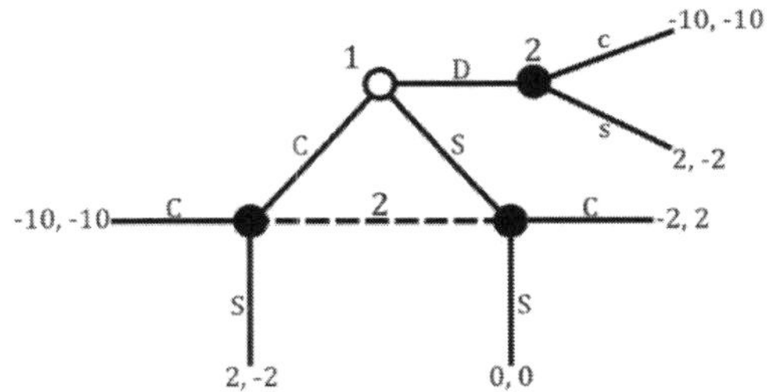

If player 1 defenestrates, player 2 has a simple decision node. Let's use backward induction there:

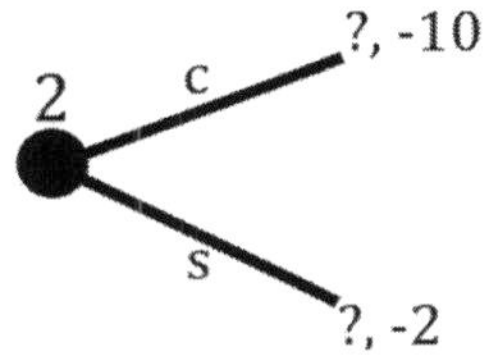

Player 2 knows player 1 must continue straight. If she also continues, she meets certain disaster and earns -10. If she swerves, she earns -2. Since -2 beats -10, she swerves.

Removing the continue strategy at that decision node leaves the following game:

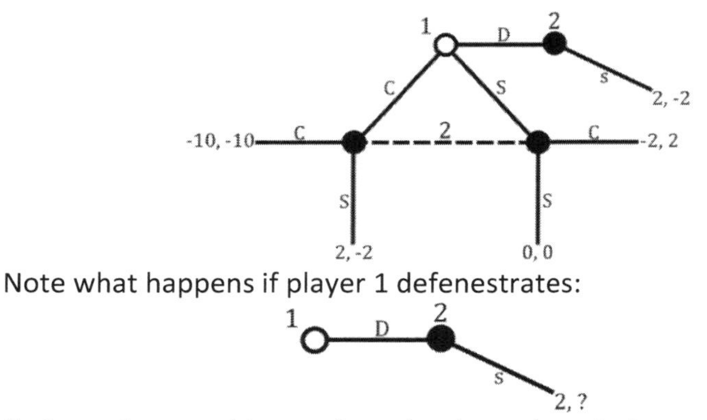

Note what happens if player 1 defenestrates:

If player 1 tosses his steering wheel out the window, player 2 knows to swerve. Player 1 earns 2 for this outcome, which is his greatest expected utility possible, which in turn means he has no profitable deviation. Player 2 cannot profitably deviate either, since she must swerve if he defenestrates. Thus, this must be an outcome that occurs in an equilibrium.

We have discussed *ad nauseum* about how subgame perfection is the study of credible threats. Defenestrated chicken teaches a valuable lesson about how to make threats credible. In the original, simultaneous move chicken, both players wanted to force the other to swerve. However, these threats were not necessarily credible; if both insisted that they would not swerve, they wind up in the fiery disaster outcome. Yet, by defenestrating his steering wheel, player 1 demonstrates an inability to swerve in the future. In turn, player 2 knows player 1's threat to not swerve is credible—he literally must continue going straight at that point. So player 1's crafty maneuver earns him his highest payoff.

Most analyses of defenestrated chicken stop there. We have certainly found *an* outcome that can occur in equilibrium, but we have not proven the uniqueness of this equilibrium outcome. We also have not found a complete equilibrium; the equilibrium does not explain what player 2 would do at her other information set, where she does not know if player 1 continued or swerved. To complete our analysis, we must address both of these issues.

Let's begin with player 2's other information set. She returns to a familiar dilemma:

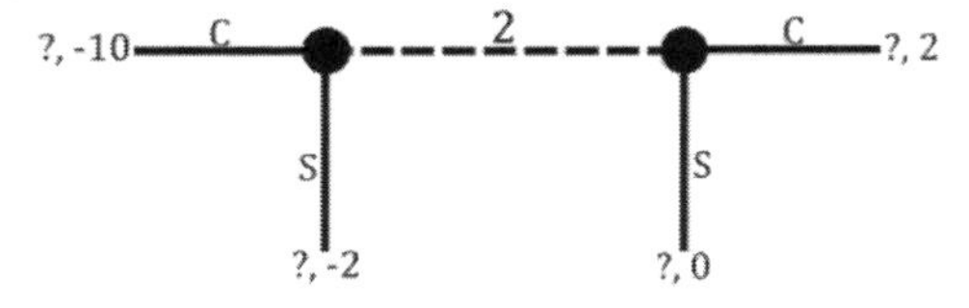

If player 1 continued, player 2 is on the left side of her information set. She should therefore swerve, as -2 beats -10. But if player 1 swerved, player 2 is on the right side of her information set. In that case, she should continue, as 2 beats 0. Thus, player 2's optimal strategy depends on what player 1 did. But unfortunately for her, she cannot see his movement.

Nevertheless, forward induction allows her to infer player 1's choice. Note that player 1 would never swerve:

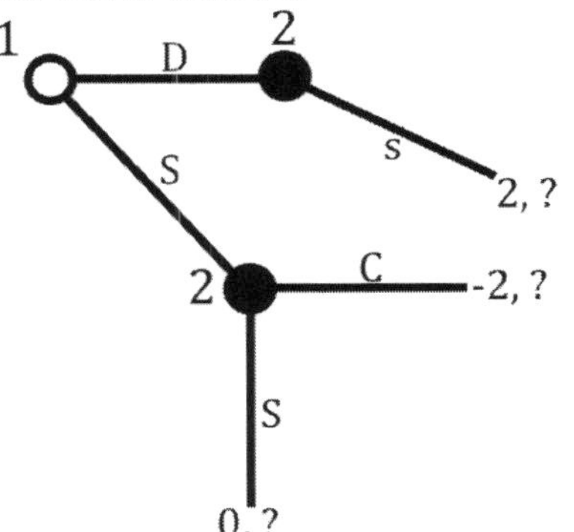

If player 1 swerves, he earns 0 or -2. However, if he defenestrates, player 2 swerves with certainty, and player 1 earns 2. Since beats both 0 and -2, it is irrational for player 1 to swerve.

With that in mind, suppose player 2 has an opportunity to move. Recall that she sees this:

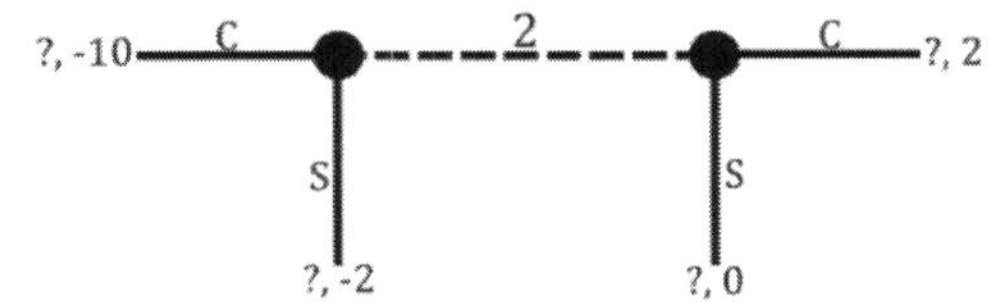

She cannot directly observe whether player 2 has continued or swerved. However, she should believe player 1's past move was rational. Since swerving is irrational for him, she therefore infers that he continued. As such, she can think of her information set as a single decision node:

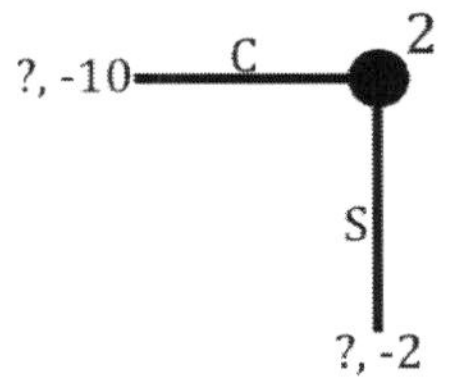

If she continues, she earns -10. If she swerves, she earns -2. Since -2 beats -10, she swerves.

Player 1 knows player 2 will infer this information, so he can view the game accordingly:

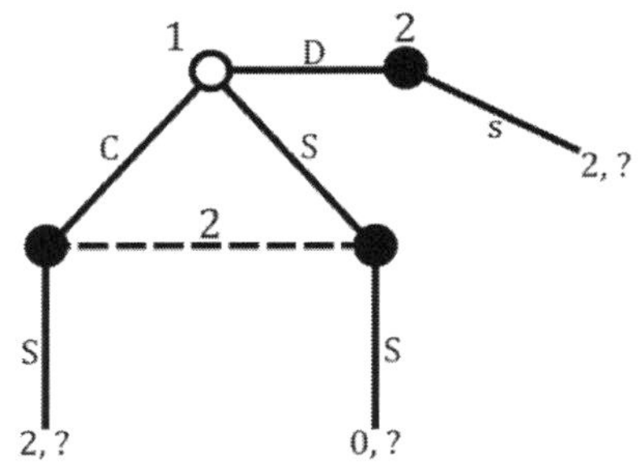

If player 1 defenestrates, player 2 swerves, and he earns 2. If he swerves, player 2 swerves, and he earns 0. Finally, if he continues, player 2 swerves, and he earns 2. Therefore, player 1 earns the most if he defenestrates or he continues. Since both of these strategies ultimately end with him earning 2, he is indifferent between them. Therefore, he can play defenestrate as a pure strategy, continue as a pure strategy, or mix freely between the two. Regardless, player 2 always swerves. Despite our best intentions, we are left with infinitely many equilibria.

As a practical matter, we may wonder whether any equilibrium in which player 1 continues with positive probability is plausible. Forward induction remains a controversial topic in game theory because it requires players to make strong inferences about the other players' behaviors. To see why, imagine player 2 believed player 1 was going to play the defenestrate equilibrium, player 1 gunned toward her full throttle without tossing his steering wheel. Although the game says that player 1's move is optimal, player 2 might imagine he is irrational or made a mistake, leading her unsure whether she should optimally continue or swerve. In turn, perhaps a rational player 1 should just defenestrate at the start rather than leaving his rationality in doubt.

Alternatively, we should simply view this result as a function of its assumptions. If both players are rational, understand backward induction, and understand forward induction, then all of those equilibria are plausible. If one of these actors is not rational, does not understand backward induction, or does not understand forward induction, then we should accept the fact that our inaccurate assumptions can lead to inaccurate predictions.

2.8.3: Costly Defenestration

Consider the same defenestrated chicken setup with a slight modification: it is costly for player 1 to defenestrate his steering wheel. Perhaps his window will not open or close, so he must shatter it to toss out the wheel. Perhaps he finds it costly to drive back to pick it up. Perhaps the impact on the ground will break the steering wheel, and player 1 will have to replace it. Regardless, let's say the cost to him is -1. Thus, if he plays defenestrate, both of his defenestration payoffs are worth 1 less than they were before.

Here is the game tree:

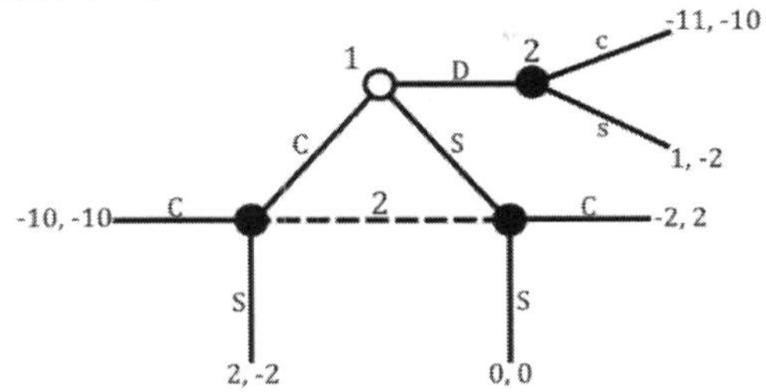

Note that the modification does not change player 2's optimal strategy if player 1 defenestrates:

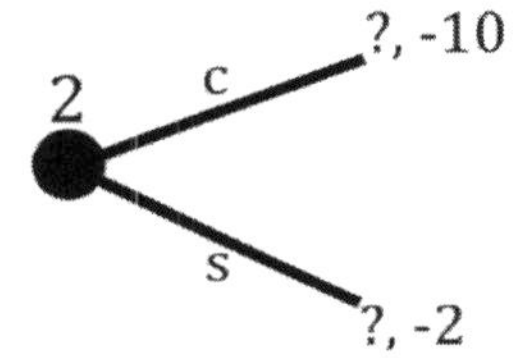

Just as before, -2 beats -10. As such, player 2 swerves.

Removing her swerve strategy from that contingency leaves us with the remaining game tree:

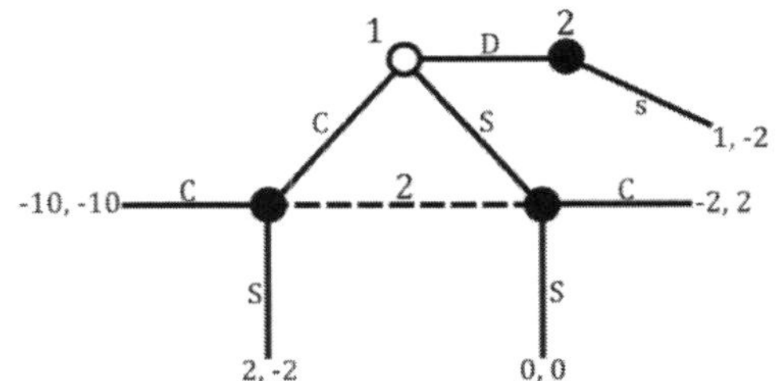

It still does not make sense for player 1 to ever swerve:

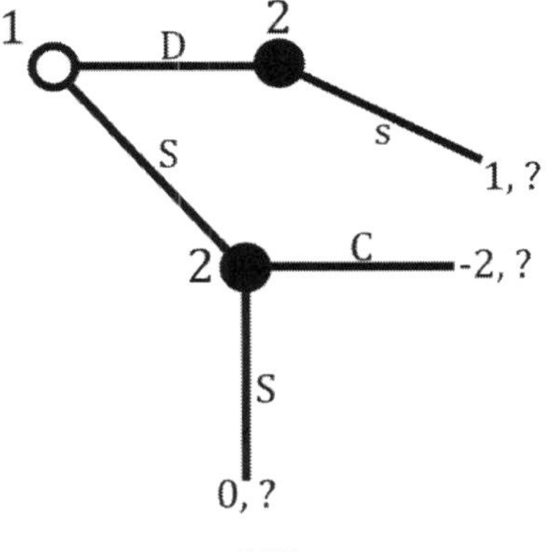

Player 1's payoff of 1 for defenestration still beats both 0 and -2. So swerving is irrational.

Now consider player 2's move:

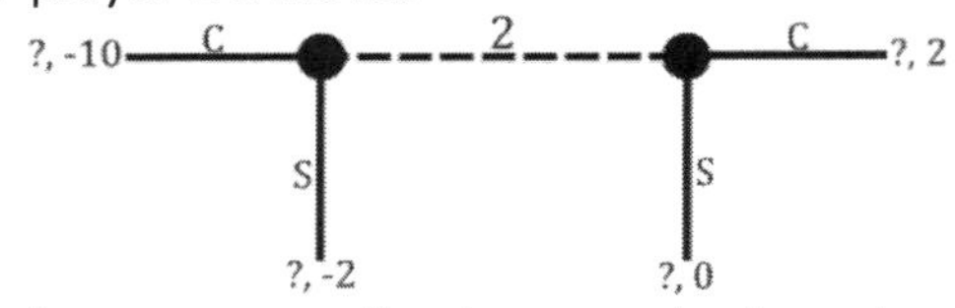

Although she does not actually observe whether player 1 has continued or swerved, forward induction allows her to infer that he continued. Thus, she only needs to consider this decision:

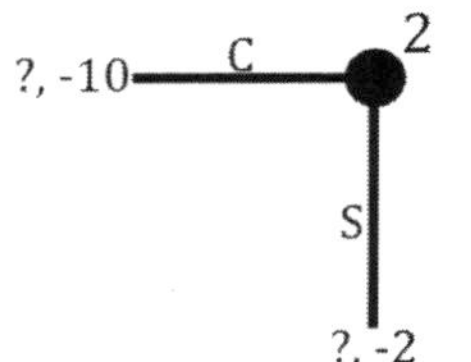

Since -2 beats -10, player 2 should swerve.

Now go back to player 1's decision, knowing that player 2 will swerve if given the opportunity to move:

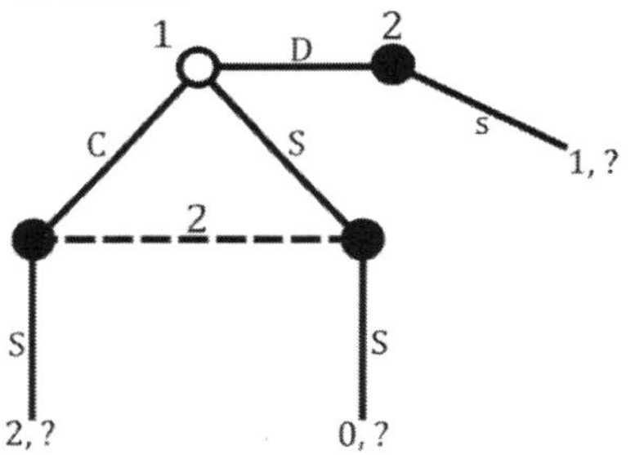

If player 1 defenestrates, player 2 swerves, and he earns 1. If he continues, she swerves, and he earns 2. If he swerves, she swerves, and he earns 0. Since 2 beats both 1 and 0, player 1 should continue. Therefore, the forward induction solution is for player 1 to continue and for player 2 to swerve at both her information sets. The first player does not even have to make a visible threat to get his way!

2.8.4: Burned Battle of the Sexes

Consider the following revised game of battle of the sexes. Player 1 begins by choosing two things. First, he decides whether to burn some money. Second, he goes to either the ballet or the fight. Player 2 observes whether player 1 burned the money or not but cannot see which venue player 1 went to. She then decides whether to go to the ballet or the fight. Each player earns 4 if they coordinate at his or her most preferred venue, 1 if they coordinate at his or her least preferred venue, and 0 if they go to different locations. In addition, if player 1 burned money, he subtracts 2 from his outcome.

Since that is a lot of information to take in, let's compress it into a game tree:

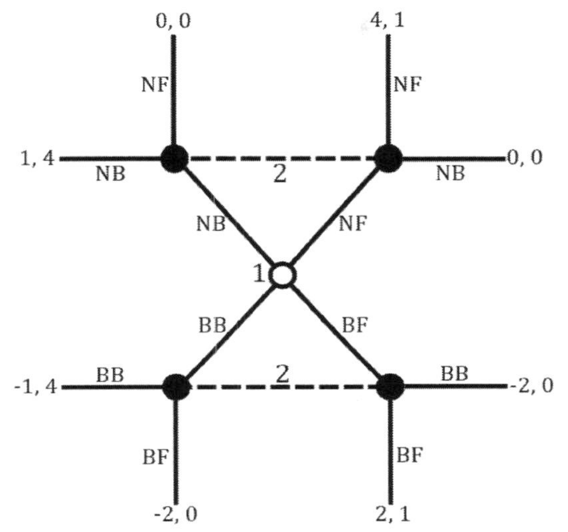

To recap, player 1 begins in the center and chooses among BB, BF, NB, and NF. The first letter refers to whether he **B**urned or did **N**ot burn the money, while the second letter is whether he goes to the **B**allet or the **F**ight. Player 2 only observes whether player 1 burned or not. If he did, she selects between BB and BF. If he did not, then she picks NB or BF.

Regardless of player 1's decision to burn, player 2 is seemingly in a quandary. Suppose player 1 does not burn. Player 2 sees this:

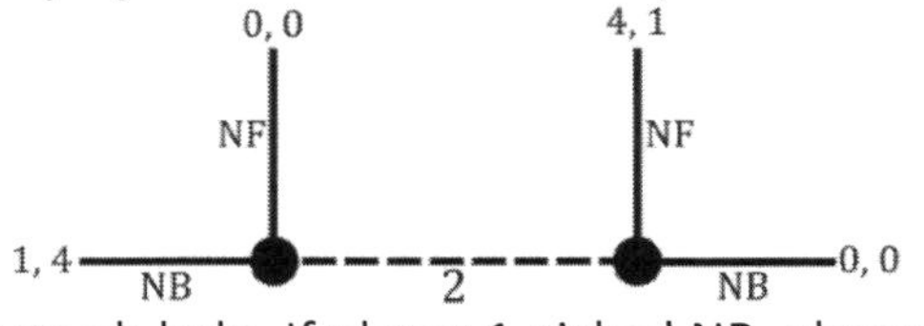

But this is not much help. If player 1 picked NB, player 2's decision is as follows:

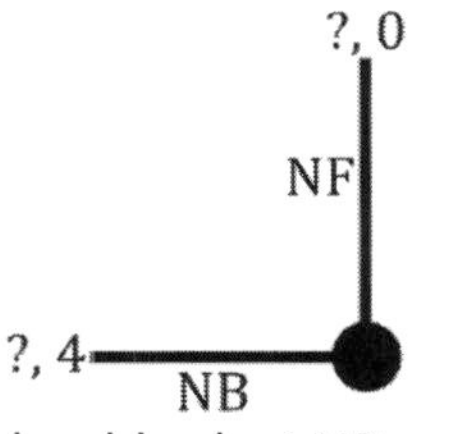

Since 4 beats 0, player 2 should select NB as well.

But if played 1 chose NF instead:

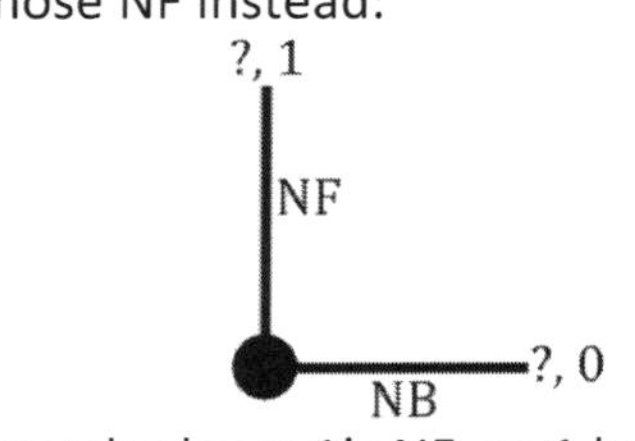

Player 2 should match player 1's NF, as 1 beats 0.

The same problem occurs if player 1 burns at the start. If he does, player 2 sees this:

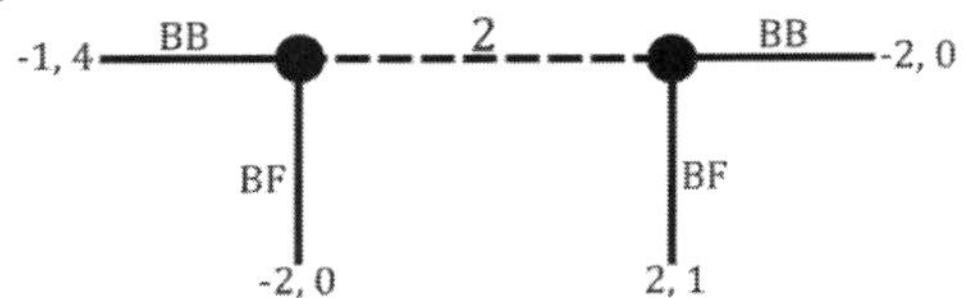

Again, this information does not help player 2. If player 1 picked BB, player 2 faces the following decision:

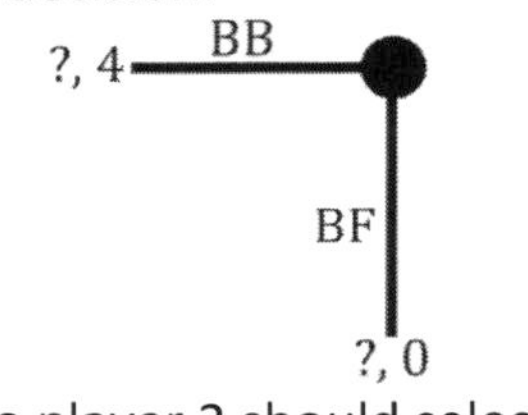

Once more, 4 beats 0, so player 2 should select BB.

But if player 1 chose BF instead:

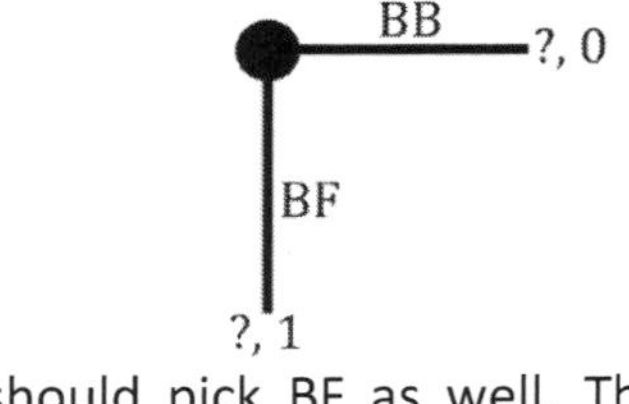

Then player 2 should pick BF as well. Thus, regardless of player 1's decision to burn, player 2 is in the familiar battle of the sexes dilemma.

Surprisingly, however, forward induction leads to a *unique* equilibrium. To unravel the twisted logic, let's begin by looking at player 1's payoffs if he chooses a non-burning strategy:

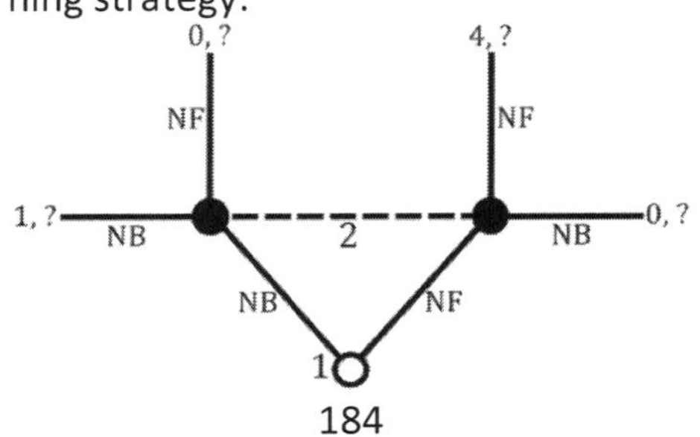

Depending on the venues the players select, player 1 can earn 0, 1, or 4. Thus, if he does not burn, his *minimum* payoff is 0; he cannot possibly earn a negative amount if he picks NB or NF at the start of the game.

Now consider player 1's payoffs if he burns at the beginning:

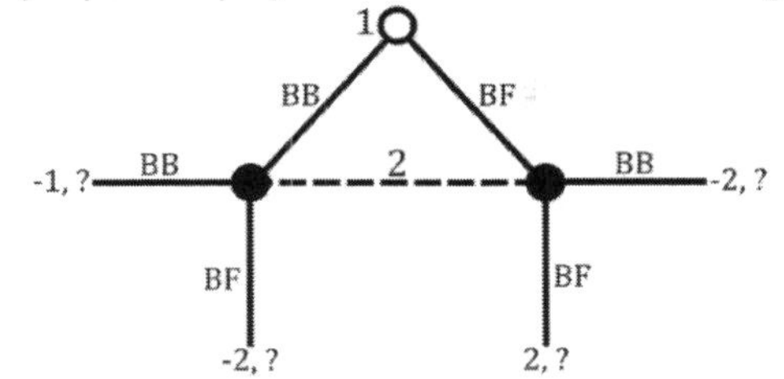

Three of these payoffs are negative. The only way he can reach a positive expected utility is if both players go to the fight.

Note that this has an interesting implication regarding his BB strategy. If player 1 selects BB at the start, he earns -1 or -2 depending on what player 2 does. Either way, he does strictly worse than if he had not burned at the beginning of the game; all of player 1's payoffs are at least 0 if he does not burn. Therefore, player 1 would never rationally play BB.

In turn, suppose player 2 had a chance to move at this information set:

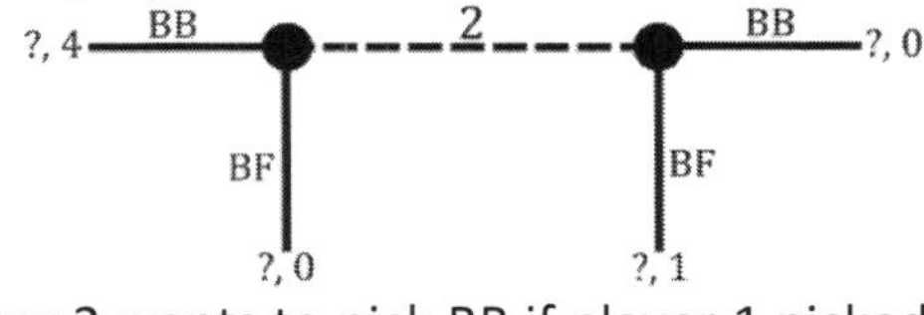

We know player 2 wants to pick BB if player 1 picked BB, but she wants to pick BF if he picked BF. Forward induction requires player 1's strategy to be rational. Since BB is an irrational strategy, player 2 can safely assume player 1 played BF:

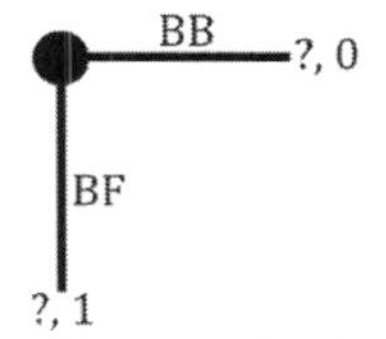

At this point, player 2 plays BF as well, since 1 beats 0.

Now consider player 1's choice if he burns:

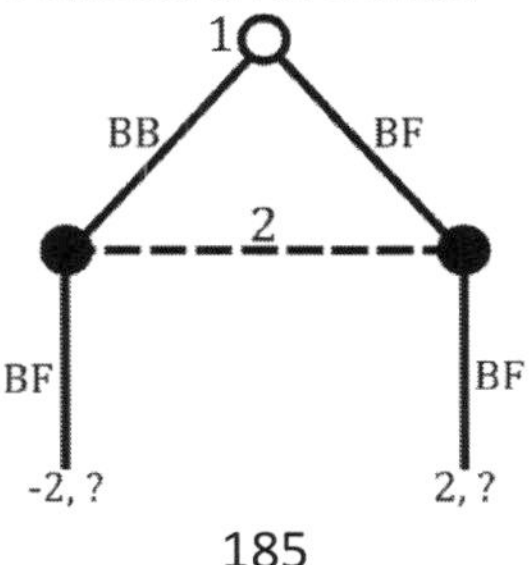

If he burns and goes to the fight, player 2 goes to the fight as well, and he earns 2. If he burns and goes to the ballet, he earns -2. As such, if he burns, he *must* go to the fight. In turn, he *must* earn at least 2 from the overall game.

However, he could potentially do better in equilibrium by not burning. With that in mind, note that not burning and going to the ballet is now an irrational strategy for player 1:

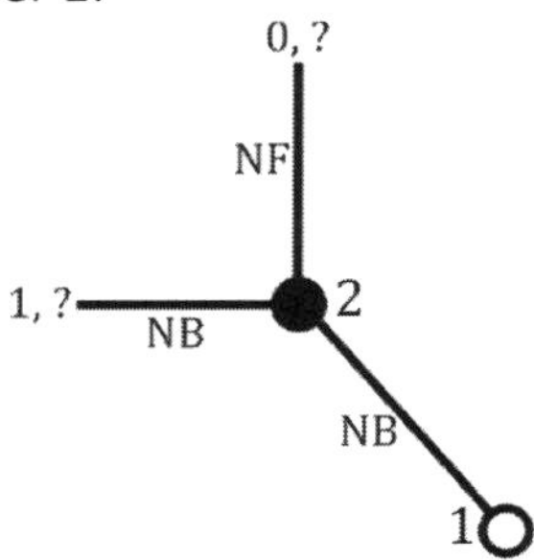

If player 1 selects NB, he ultimately earns 1 or 0. Both of these payoffs are less than the 2 that he could earn from BF. Therefore, player 1 cannot play NB.

Given that, suppose player 2 reaches this information set:

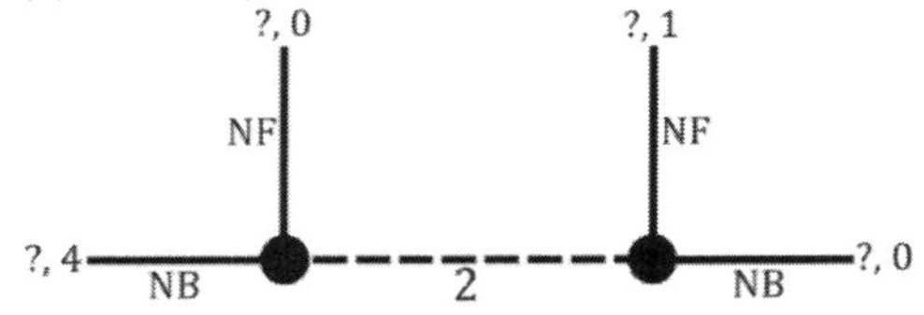

Player 2 cannot see whether player 1 selected NB or NF. However, through the long line of above reasoning, player 2 can infer that player 1 would never play NB. As such, if she is at this information set, she knows player 1 must have played NF. Therefore, she can narrow her decision down to this:

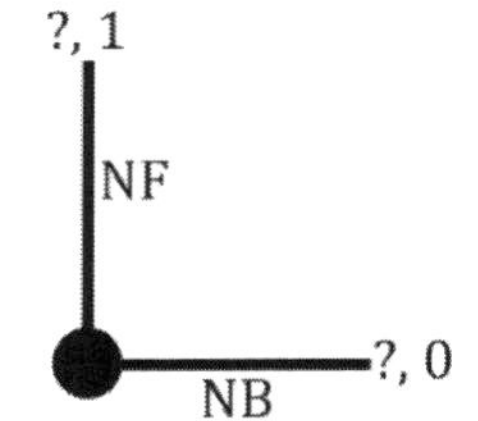

If player 2 plays NF, she earns 1; if she plays NB, she earns 0. Since 1 beats 0, she chooses NF.

Finally, consider player 1's optimal decision at the beginning of the game. Regardless of his decision to burn money, he knows player 2 will go to the fight. Therefore, he faces the following decision:

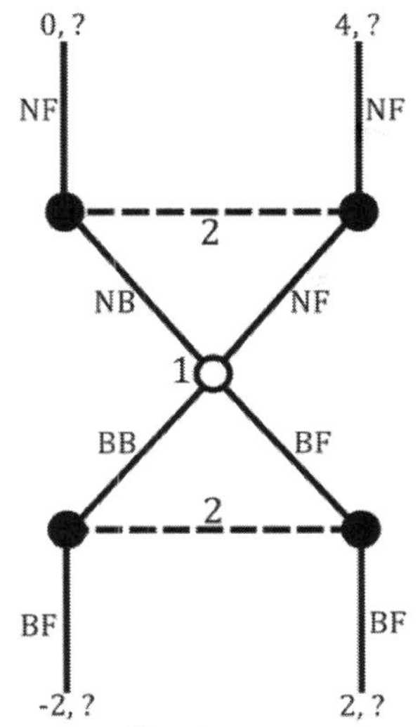

Since 4 is the greatest payoff, player 1 selects NF. Thus, the ability to burn money delivers player 1 his best outcome even though he never actually burns any in equilibrium.

Takeaway Points

1) Forward induction assumes all past play was rational.
2) The inferences players make using forward induction require extremely sophisticated thinking.

Lesson 3.1: Probability Distributions

In this chapter, we will investigate matrix games far more complex than what we saw in Chapter 1. There will be two new levels of difficulty. First, we will abandon specific numerical payoffs to generalize our games as much as possible. For instance, consider the strategic interaction between a striker and a goalie during a soccer penalty kick. The interaction is essentially a guessing game; the striker does not want the goalie to dive in the direction of the kick, but the goalie does. Suppose the striker has perfect aim to the right side but only shoots on target on the left side with probability x, where $0 < x < 1$. Then the matrix looks like this:

	D Left	D Right
K Left	0, 0	x, -x
K Right	1, -1	0, 0

We have never encountered a variable inside of a payoff matrix before. At present, we cannot solve this game.

In this chapter, we will also see games that have mixed strategy Nash equilibria utilizing more than two pure strategies. Rock-paper-scissors is an example:

	Rock	Paper	Scissors
Rock	-3, 6	9, 1	0, 2
Paper	3, -4	2, 4	4, 1
Scissors	4, 7	3, 2	-3, 2

If either player develops a pattern in rock-paper-scissors, his or her opponent can exploit it. Consequently, in equilibrium, the players must randomize among all three strategies. However, we only know how to solve for mixed strategies involving two pure strategies. Again, at present, we are helpless here.

Over the course of this chapter, we will frequently work with probabilities in the form of mixed strategies. These probabilities will not be as friendly as before, as they will often span three strategies or include variables. For example, the probability $x/(x + y + z)$ is meaningless at the moment but will eventually become second nature.

In turn, we must have a decent understanding of probability theory before moving on. To that end, this lesson teaches two things. Primarily, we will learn about probability distributions—what they are, where we have seen them before, and why we need to know about them. Second, we will create a way to test whether numerical expressions form a valid probability distribution. If they do, then we can use those probabilities for our mixed strategies. If not, we can dismiss certain potential mixed strategy Nash equilibria.

3.1.1: The Golden Rules of Probability Distributions

A probability distribution is a set of events and the probability each event in the set occurs. For example, when I flip a coin, the probability it lands on heads is 1/2 and the probability it lands on tails is also 1/2. When I roll a die, the probability it lands on 1 is 1/6, the probability it lands on 2 is 1/6, the probability it lands on 3 is 1/6, and so forth. When I spin a roulette wheel, the probability the ball stops on a red space is 18/38, the probability it stops on a black space is 18/38, and the probability it lands on a green space is 2/38.

The past few sentences simply linked events (the side of the coin landing upward, the side of the die coming on top, the ball stopping on a particular color) to their respective probabilities. Probability distributions are that simple.

Two golden rules of probability distributions maintain their mathematical tractability:

1) All events occur with probability no less than 0.
2) The sum of all probabilities of all events equals 1.

Although these rules are basic, four implications follow from them. First, no probability can be greater than 1. If this were the case, then there would have to be some events that occur with negative probability for all of the probabilities to sum to 1. However, such an event violates the first golden rule that requires all probabilities to be at least 0. Substantively, this implication makes perfect sense. After all, some event cannot occur more than 100% of the time.

Second, probability distributions cannot leave us wondering what else might happen. For example, a probability distribution cannot only say that the world will end tomorrow with probability 1/100. Such a distribution does not sum to 1. Perhaps the person who wrote the distribution meant that the world will not end with probability 99/100, but we need to know that explicitly.

Game theorists sometimes play loosely with this rule. Mixed strategies are probability distributions. They associate events (the pure strategies a player might choose) with a probability. When writing equilibria, we will often state something like "player 1 plays heads with probability 1/3 and player 2 plays left with probability 4/5." Technically, these are not valid probability distributions. What does player 1 do the remaining 2/3 of the time? What does player 2 do the remaining 1/5 of the time?

Implicitly, we might mean that player 1 plays down and player 2 plays right with the remaining probabilities. While imprecise, the shortcut is serviceable in the realm of simple 2x2 matrix games. However, in this chapter, we will consider games where players might want to mix among more than two strategies. If player 1 can select from up, middle, or down, writing "player 1 selects up with probability 1/3" is insufficient. We must know what percentage of the time he plays middle and what percentage of the time he plays down.

Third, the fact that the sum of the probabilities of all events equals 1 gives us a convenient way to solve for an unknown probability. Suppose a probability distribution is valid, but we do not know the probability of one of the events. Since the probabilities must sum to 1, the remaining probability equals 1 minus the sum of the probabilities of the rest of the events. This is why we only solve for σ_{up} rather than σ_{up} and σ_{down} simultaneously in a 2x2 game; we know mixed strategies form valid probability distributions, so σ_{down} must be equal to $1 - \sigma_{up}$. Again, while this is simple in the two strategy case, we will be working with more difficult examples in this chapter involving three strategies.

Fourth, probabilities can be 0 or 1. An event with probability 0 never occurs, while an event with probability 1 always occurs. In fact, on a technical level, pure strategies are *mixed* strategies because of this feature. The probability distribution for the "mixture" here is 1 on the pure strategy played in equilibrium and 0 on all other strategies.

3.1.2: Testing the Validity of a Probability

Consider the following game:

	Left	Right
Up	R, r	S, t
Down	T, s	P, p

This is a generalized prisoner's dilemma. We will formally solve it in Lesson 3.2. Ultimately, we want to show whether a potential mixed strategy Nash equilibrium is valid. To do this, we must solve for mixtures that are an amalgamation of letters rather than numbers. But to confirm the integrity of our solutions, we need to show that the probabilities form a valid probability distribution.

For example, consider the fraction x/y, where x and y could potentially be any real number. Let's think about circumstances that would show a violation of the rules of probability distributions.

First, y cannot equal 0. If it did, then x/y would be undefined. We cannot sum an undefined element, so any such probability is invalid.

Can x be 0? Sure, as long as y is not as well. Any fraction 0/y reduces to just 0. The first of the golden rules of probability distributions allowed for events to occur with probability 0. So the probability is legitimate if trivial.

Next, if x is not equal to 0, x and y must both be positive or both be negative. If one is positive and the other is negative, then we have a negative probability. But probabilities cannot be negative, so we would have to reject such a figure. In contrast, a positive number divided by a positive number yields a positive number. A negative number divided by a negative number cancels out the negations and yields a positive number as well. Thus, these last two cases are permissible.

Finally, the magnitude of x must be less than or equal to the magnitude of y. For example, suppose both x and y were positive. If x were greater than y, then the probability would be greater than 1. But probabilities cannot be greater than 1, so this would be invalid.

This rule is a little more tricky if x and y are both negative. Since we care about magnitude here, x must be *greater* than or equal to y for the probability to be valid. Take, for example, -1/-3. Clearly this is a valid probability distribution because it reduces to 1/3, even though -1 is greater than -3.

Thus, we must be extra careful with this step and employ the appropriate test. If x and y are both positive, we must verify that y ≥ x. But if x and y are both negative, we need to ensure x ≥ y. We must not mix up these tests, or we usually draw an incorrect conclusion.

In any case, we now know how to test whether a figure can be a part of a valid probability distribution. The process is merely three steps:

1) If denominator is 0, the probability is invalid. If it is not, advance to the second step.
2) If the numerator is zero, the probability is valid, and the third step is irrelevant. If it is positive and the denominator is positive, advance to the third step. If it is negative and the denominator is negative, advance to the third step. If one is positive and the other is negative, the probability is invalid.

3) If the absolute value of the numerator is less than or equal to the absolute value of the denominator, the probability is valid. Otherwise, it is not.

The next two sections will put this plan into action repeatedly. If these rules about probability distributions are unclear now, you will have plenty of practice ahead.

Takeaway Points

1) A *probability distribution* is a set of events and the associated probability that each even occurs.
2) Mixed strategies are probability distributions over pure strategies.
3) All probabilities in a probability distribution cannot be less than zero and must sum to 1.

Lesson 3.2: Mixed Strategy Nash Equilibria in Generalized Games

This lesson offers our first look at generalize games. We will start off easy by analyzing battle of the sexes. Everything behaves nicely in that game—the pure strategy Nash equilibria are easy to find, and the mixed strategy algorithm produces a sensible result. However, we will then cover the prisoner's dilemma and deadlock to see how to derive contradictions when we replace numbers with variables.

3.2.1: Generalized Battle of the Sexes

In Lesson 1.6, we solved the following game of battle of the sexes:

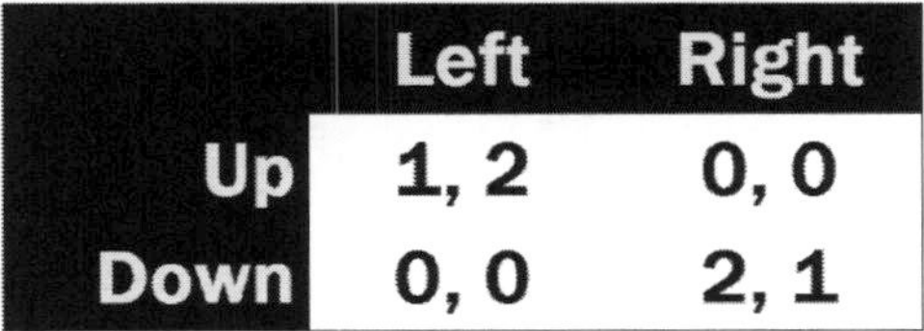

	Left	Right
Up	1, 2	0, 0
Down	0, 0	2, 1

Battle of the sexes has two pure strategy Nash equilibria, <up, left> and <down, right>, and a mixed strategy Nash equilibrium in which player 1 played up with probability 1/3 and player 2 played left with probability 2/3.

In Lesson 2.8, we solved an extensive form game of battle of the sexes in which player 1 could burn money before making a play. However, the payoffs were slightly adjusted—a player who went to his or her more preferred form of entertainment and met his or her partner there earned 4 instead of 2. If we take away the burning money aspect of the game and convert the extensive form to a matrix, we arrive at this:

	Left	Right
Up	1, 4	0, 0
Down	0, 0	4, 1

The modified game has the same look and feel as the original battle of the sexes. The players still earn 0 if they fail to coordinate, but one each one prefers going to a different location.

It should be obvious that <up, left> and <down, right> are still pure strategy Nash equilibria as well. However, the mixed strategy Nash equilibrium is different. Let's solve for player 1's mixed strategy that leaves player 2 indifferent between left and right:

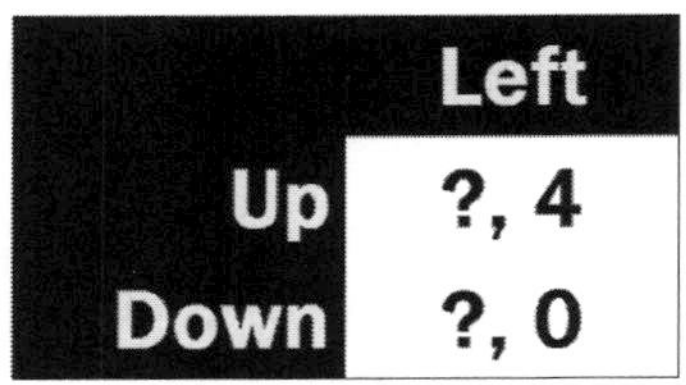

	Left
Up	?, 4
Down	?, 0

If player 2 moves left, she earns 4 with probability σ_{up} and 0 with probability $1 - \sigma_{up}$. As an equation:

$EU_{left} = (\sigma_{up})(4) + (1 - \sigma_{up})(0)$

Let's check right:

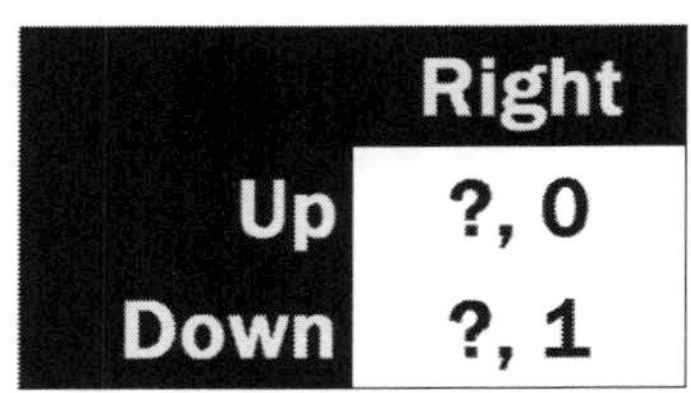

	Right
Up	?, 0
Down	?, 1

If player 2 moves right, she earns 0 with probability σ_{up} and 1 with probability $1 - \sigma_{up}$. As an equation:

$EU_{right} = (\sigma_{up})(0) + (1 - \sigma_{up})(1)$

Now we set her expected utility for left equal to her expected utility for right and solve for σ_{up}:

$EU_{left} = EU_{right}$
$EU_{left} = (\sigma_{up})(4) + (1 - \sigma_{up})(0)$
$EU_{right} = (\sigma_{up})(0) + (1 - \sigma_{up})(1)$
$(\sigma_{up})(4) + (1 - \sigma_{up})(0) = (\sigma_{up})(0) + (1 - \sigma_{up})(1)$
$4\sigma_{up} = 1 - \sigma_{up}$
$5\sigma_{up} = 1$
$\sigma_{up} = 1/5$

So in the mixed strategy Nash equilibrium, player 1 moves up with probability 1/5 and down with probability 4/5.

Let's move on to player 2's mixed strategy:

	Left	Right
Up	1, ?	0, ?

If player 1 moves up, he earns 1 with probability σ_{left} and 0 with probability $1 - \sigma_{left}$. As an equation:

$$EU_{up} = (\sigma_{left})(1) + (1 - \sigma_{left})(0)$$

Now for down:

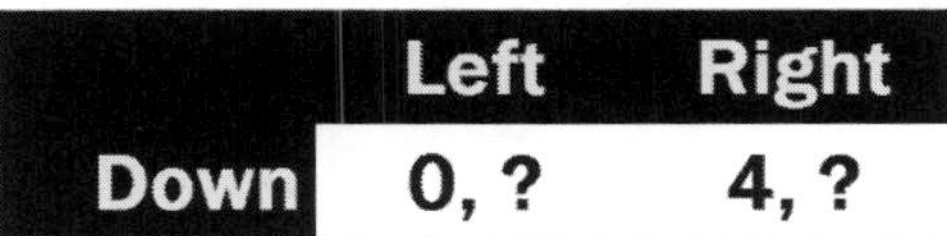

	Left	Right
Down	0, ?	4, ?

If player 1 selects down, he earns 0 with probability σ_{left} and 4 with probability $1 - \sigma_{left}$. As an equation:

$$EU_{down} = (\sigma_{left})(0) + (1 - \sigma_{left})(4)$$

Once again, to find player 2's mixed strategy that leaves player 1 indifferent between up and down, we set these two expected utilities equal to each other and solve for σ_{left}:

$$EU_{up} = EU_{down}$$
$$EU_{up} = (\sigma_{left})(1) + (1 - \sigma_{left})(0)$$
$$EU_{down} = (\sigma_{left})(0) + (1 - \sigma_{left})(4)$$
$$(\sigma_{left})(1) + (1 - \sigma_{left})(0) = (\sigma_{left})(0) + (1 - \sigma_{left})(4)$$
$$\sigma_{left} = 4 - 4\sigma_{left}$$
$$5\sigma_{left} = 4$$
$$\sigma_{left} = 4/5$$

So in the mixed strategy Nash equilibrium, player 2 goes left with probability 4/5 and down with probability 1/5. Thus, increasing the value of the most preferred outcome changes the mixed strategy Nash equilibrium.

Notice that we can make infinitely many of such changes to the game. Instead of a 2 or a 4 for the most preferred outcome, we could switch it to 1.5, 3, π, 13, or 100. Alternatively, we change the payoff for not coordinating from 0 to -4, -8, -15, -16, -23, or -42. Every time we change the payoffs, no matter how trivially small the differences are, we must recalculate the mixed strategy Nash equilibrium.

If we are going to encounter many different versions of battle of the sexes, it would help if we could derive a simple formula for the mixed strategy Nash equilibrium. At present, we only have the mixed strategy algorithm. That algorithm eventually finds the MSNE, but it requires a burdensome number of calculations each time we run it. Ideally, we would like to make the algorithmic calculations once and be able to apply the results every time we encounter an altered version of the game.

Fortunately, we can make such an abstraction. All we have to do is replace the distinct numbers with *exogenous variables*:

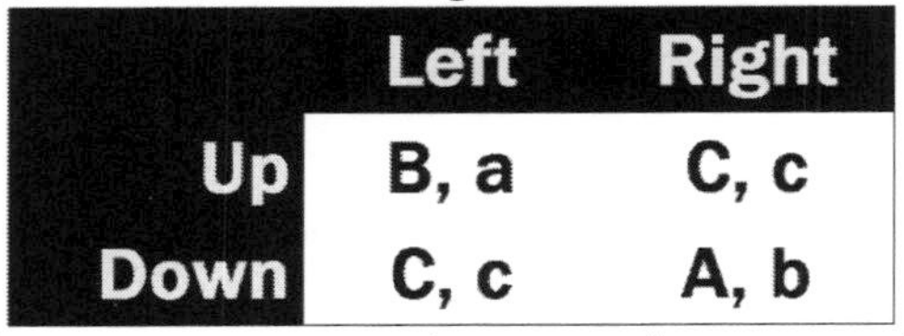

	Left	Right
Up	B, a	C, c
Down	C, c	A, b

These variables are exogenous because they come from outside the game. The players do not choose them. (If they players chose the values, they would be *endogenous variables*.) Instead, the players have these payoffs and know them going into the game, as though they were innate preferences.

Note that if we create a rule that A > B > C and a > b > c, we have the correct preference ordering for battle of the sexes. In the original game, A = a = 2, B = b = 1, and C = c = 0. In the modified version, we changed A and a to 4. Regardless of the specific values, the players are trying to coordinate at a specific location, though their most preferred outcomes differ.

Another useful feature of this general form is that we can easily consider different payoffs for each player. Thus, player 1's payoffs are in capital letters while player 2's payoffs are in lower case. These payoffs may be the same as they were in the first two examples, but they can also be different. For instance, player 1 could earn 4 for coordinating at the fight (as in the burned version) while player 2 might earn 2 for coordinating at the ballet (as in the original version). As it turns out, these differences do not make our generalized analysis any more difficult.

Let's solve the game. The process is the same as before. The only extra difficulty is that we have variables to work with instead of actual numbers. This will be a slight inconvenience to us. However, as long as we understand that these variables behave exactly how ordinary numbers do, we will be fine.

First, let's mark the best responses of the game, beginning with how player 1 should respond to player 2 selecting left:

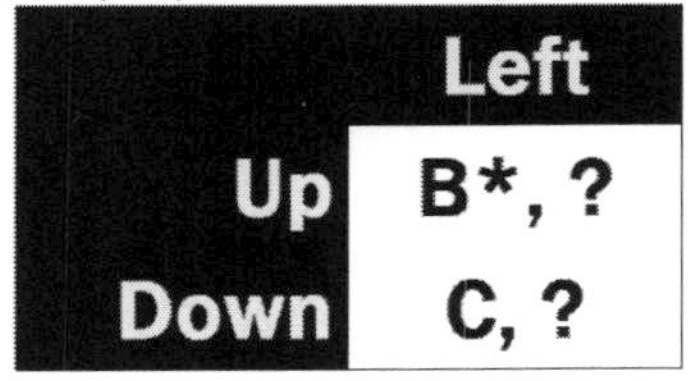

	Left
Up	B*, ?
Down	C, ?

Recall that B > C. Therefore, player 1's best response is to play up, so B earns the asterisk.

Now for right:

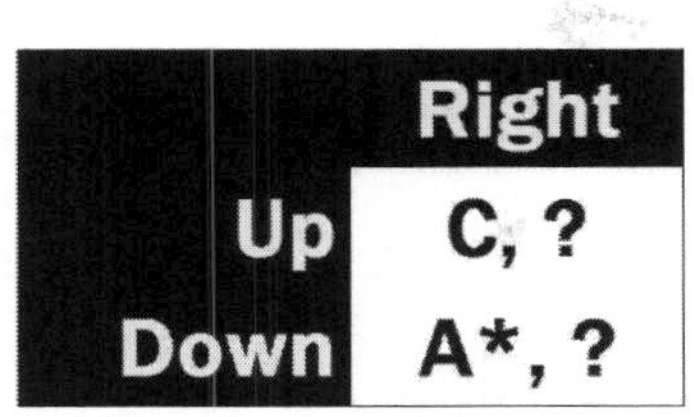

	Right
Up	C, ?
Down	A*, ?

A beats C. Therefore, A has the asterisk.

Switching gears, let's find player 2's best responses, beginning with player 1 choosing up:

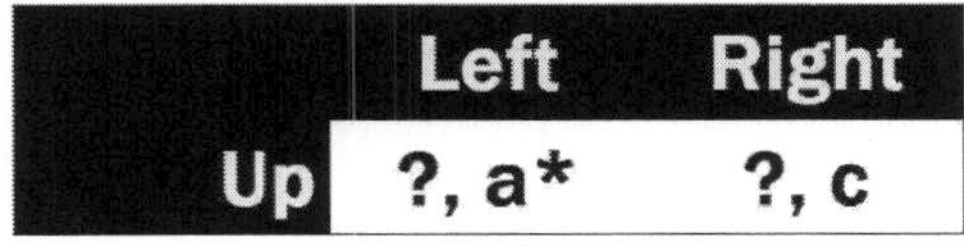

	Left	Right
Up	?, a*	?, c

Once again, a beats c. As such, left is player 2's best response, and it receives an asterisk.

Now for down:

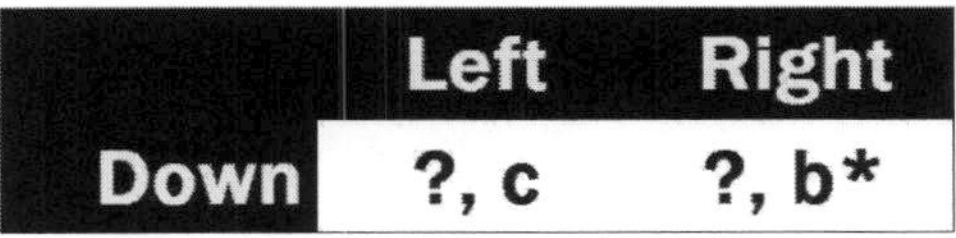

	Left	Right
Down	?, c	?, b*

This time, b earns the asterisk, as b > c.

Putting all of these best responses together, we arrive at the following:

	Left	Right
Up	B*, a*	C, c
Down	C, c	A*, b*

Both <up, left> and <down, right> are mutual best responses, so they are pure strategy Nash equilibria. Despite the variables, we found PSNE using the exact same method as before.

We still need to check for MSNE. Solving for mixed strategies is more difficult than finding PSNE, but the process is manageable. Let's start with player 1's mixed strategy that leaves player 2 indifferent between left and right. First, we find her expected utility for left:

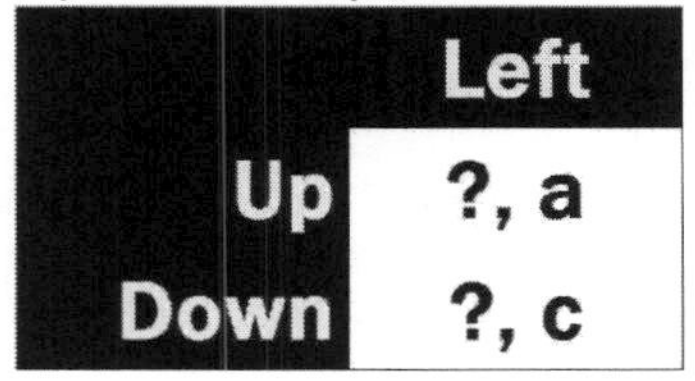

	Left
Up	?, a
Down	?, c

If player 2 plays left, she earns a with probability σ_{up} and c with probability $1 - \sigma_{up}$. As an equation:

$$EU_{left} = (\sigma_{up})(a) + (1 - \sigma_{up})(c)$$

Now for right:

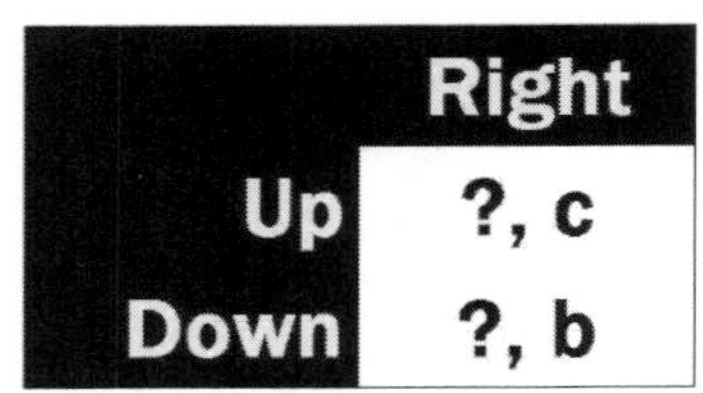

	Right
Up	?, c
Down	?, b

This time, player 2 earns c with probability σ_{up} and b with probability $1 - \sigma_{up}$. As an equation:

$$EU_{right} = (\sigma_{up})(c) + (1 - \sigma_{up})(b)$$

Just as before, we set player 2's expected utility for left equal to her expected utility for right and solve for σ_{up}:

$$EU_{left} = EU_{right}$$
$$EU_{left} = (\sigma_{up})(a) + (1 - \sigma_{up})(c)$$
$$EU_{right} = (\sigma_{up})(c) + (1 - \sigma_{up})(b)$$
$$(\sigma_{up})(a) + (1 - \sigma_{up})(c) = (\sigma_{up})(c) + (1 - \sigma_{up})(b)$$
$$a\sigma_{up} + c - c\sigma_{up} = c\sigma_{up} + b - b\sigma_{up}$$

The next steps can be confusing. We must bunch all of the terms with σ_{up} on one side of the equation and all those without on the other side:

$$a\sigma_{up} + c - c\sigma_{up} = c\sigma_{up} + b - b\sigma_{up}$$
$$a\sigma_{up} + b\sigma_{up} - 2c\sigma_{up} = b - c$$

Now we can use the distributive property of multiplication to take out each instance of σ_{up} from everything on the left side of the equation:

$$a\sigma_{up} + b\sigma_{up} - 2c\sigma_{up} = b - c$$
$$\sigma_{up}(a + b - 2c) = b - c$$

To solve for σ_{up}, we only need to divide both sides by $a + b - 2c$. However, there is a caveat here: we cannot divide by zero. As such, if $a + b - 2c = 0$, we cannot complete this task.

Fortunately, $a + b - 2c$ cannot equal zero in this instance. In fact, it must be greater than zero:

$$a + b - 2c > 0$$
$$a + b - c - c > 0$$
$$a - c > c - b$$

Since $a > c$, $a - c$ is positive. On the other hand, since $c < b$, $c - b$ is negative. A positive number is always greater than a negative number, so the inequality holds. Therefore, we may divide by $a + b - 2c$:

$$\sigma_{up}(a + b - 2c) = b - c$$
$$\sigma_{up} = (b - c)/(a + b - 2c)$$

Before concluding that player 1 plays up with probability $(b - c)/(a + b - 2c)$ in the MSNE, we must confirm the validity of the probability distribution. Specifically, we must check whether $(b - c)/(a + b - 2c)$ is between 0 and 1. The previous lesson explained the three step process; now we must implement it.

First, we need to know whether the denominator is zero. Here, the denominator is $a + b - 2c$. But when we divided by that number, we already showed that it is greater than zero. So we are covered for this part.

The second step is to verify that the numerator and denominator are both positive or both negative. Since we know the denominator is positive, we need to check whether the numerator is as well:

$$b - c > 0$$
$$b > c$$

By assumption, b is greater than c, so $b - c$ is positive.

Finally, we must check whether the numerator is less than the denominator:

$$b - c < a + b - 2c$$
$$0 < a - c$$
$$c < a$$

Since c is less than a by definition, the inequality holds. The numerator is less than the denominator.

The three-step process checked out, so $\sigma_{up} = (b - c)/(a + b - 2c)$ is a valid probability. Knowing that the probability distribution must sum to 1, we can easily solve for the probability that player 1 selects down:

$\sigma_{down} = 1 - \sigma_{up}$
$\sigma_{up} = (b - c)/(a + b - 2c)$
$\sigma_{down} = 1 - (b - c)/(a + b - 2c)$
$\sigma_{down} = (a + b - 2c)/(a + b - 2c) - (b - c)/(a + b - 2c)$
$\sigma_{down} = (a - c)/(a + b - 2c)$

We need not verify the validity of σ_{down}; the fact that σ_{up} is a valid probability and $\sigma_{down} = 1 - \sigma_{up}$ preserves the validity of σ_{down}. Together, they form a valid probability distribution.

Moving on, we must to repeat this process for player 2's mixed strategy. Let's start with player 1's expected utility for up:

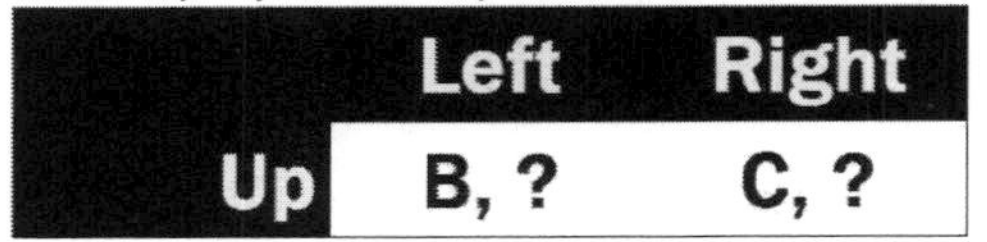

	Left	Right
Up	B, ?	C, ?

Player 1 earns B with probability σ_{left} and C with probability $1 - \sigma_{left}$. As an equation:

$EU_{up} = (B)(\sigma_{left}) + (C)(1 - \sigma_{left})$

Now for down:

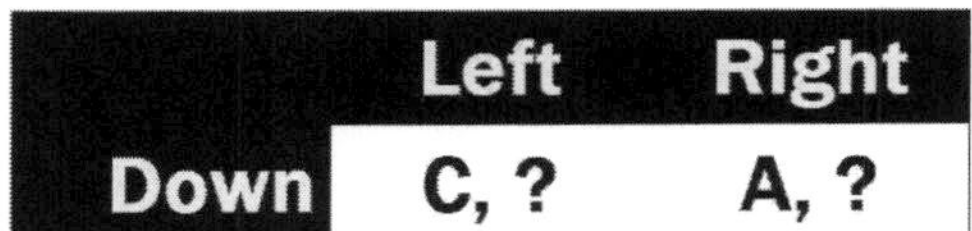

	Left	Right
Down	C, ?	A, ?

Player 1 earns C with probability σ_{left} and A with probability $1 - \sigma_{left}$. As an equation:

$EU_{down} = (C)(\sigma_{left}) + (A)(1 - \sigma_{left})$

We set player 1's expected utility for up equal to his expected utility for down and solve for σ_{left}:

$EU_{up} = EU_{down}$
$EU_{up} = (B)(\sigma_{left}) + (C)(1 - \sigma_{left})$
$EU_{down} = (C)(\sigma_{left}) + (A)(1 - \sigma_{left})$
$(B)(\sigma_{left}) + (C)(1 - \sigma_{left}) = (C)(\sigma_{left}) + (A)(1 - \sigma_{left})$

$B\sigma_{left} + C - C\sigma_{left} = C\sigma_{left} + A - A\sigma_{left}$

As before, we group all terms with σ_{left} in them on one side and the rest on the other:

$B\sigma_{left} + C - C\sigma_{left} = C\sigma_{left} + A - A\sigma_{left}$
$A\sigma_{left} + B\sigma_{left} - 2C\sigma_{left} = A - C$

Then we pull out the σ_{left} from each term on the left side of the equation:

$A\sigma_{left} + B\sigma_{left} - 2C\sigma_{left} = A - C$
$\sigma_{left}(A + B - 2C) = A - C$

Can we divide by $A + B - 2C$? Well, for the same reason that $a + b - 2c$ is greater than zero, we know that $A + B - 2C$ is also greater than zero. So we may divide as normal:

$\sigma_{left}(A + B - 2C) = A - C$
$\sigma_{left} = (A - C)/(A + B - 2C)$

Again, we must check whether $(A - C)/(A + B - 2C)$ is a valid probability distribution. We know the denominator is positive from when we divided by $A + B - 2C$, so the next step is to check whether the numerator is also positive:

$A - C > 0$
$A > C$

Since $A > C$ by definition, this holds.

Finally, we must ensure the numerator is less than the denominator:

$A - C < A + B - 2C$
$0 < B - C$
$C < B$

C is less than B by definition, so that works out. The mixed strategy forms a valid probability distribution.

Battle of the sexes demonstrated how to solve for general mixed strategy Nash equilibria when everything goes right. In practice, things often go wrong. The next two sections show how to spot some of those problems and explain why the mixed strategy algorithm does not always work perfectly.

3.2.2: Generalized Prisoner's Dilemma

Recall back to our very first lesson on strict dominance, which featured the prisoner's dilemma and deadlock. Neither of these games had a mixed strategy Nash equilibrium. Will this remain true when we generalize the game? How will we know that a MSNE does not exist when we are not even working with numbers? This lesson contains the answers: there still will not be any MSNE in the generalized games, and we can prove this using our knowledge of probability distributions.

Let's start with the prisoner's dilemma:

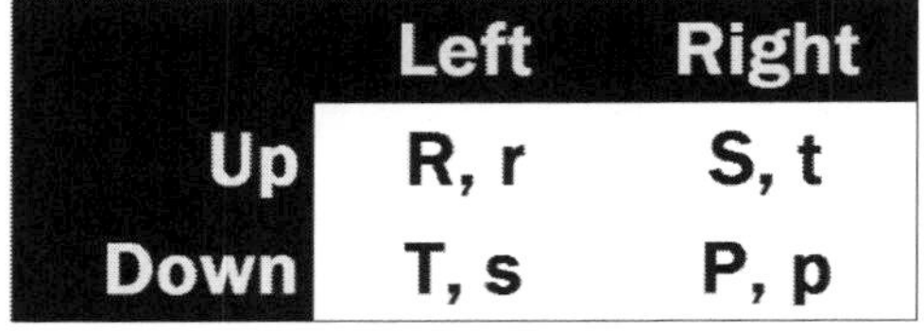

	Left	Right
Up	R, r	S, t
Down	T, s	P, p

This time around, T > R > P > S and t > r > p > s. You can remember the ordering like this: T is for **T**emptation, which is the payoff a player receives when he rats out his opponent and the opponent remains silent; R is for **R**eward, which is the good payoff the players receive when they both remain silent; P is for **P**unishment, which is the bad payoff both players receive when they both rat out the other; and S is for **S**ucker, which is the payoff for a player when he remains silent and the opponent rats him out. (These naming conventions come from *The Evolution of Cooperation* by Robert Axelrod, which is the seminal book on the prisoner's dilemma. It is extremely accessible to readers new to game theory. As such, I give it my highest recommendation.)

Just as before, we can see that up and left (the strategies that represent staying silent) are strictly dominated. Let's show this for player 1. Suppose player 2 moved left:

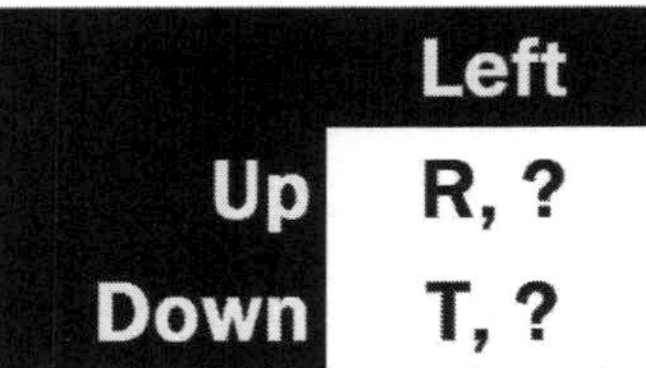

	Left
Up	R, ?
Down	T, ?

Since T beats R, player 1 should play down in this contingency.

Now suppose player 2 played right:

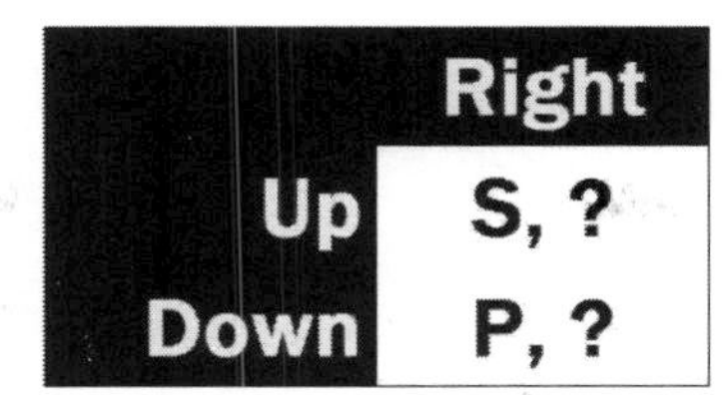

	Right
Up	S, ?
Down	P, ?

Again, player 1 ought to select down, as P beats S. Since down provides a greater payoff than up regardless of player 2's strategy, down strictly dominates up.

Let's switch perspectives and look at player 2's decision. Suppose player 1 chose up:

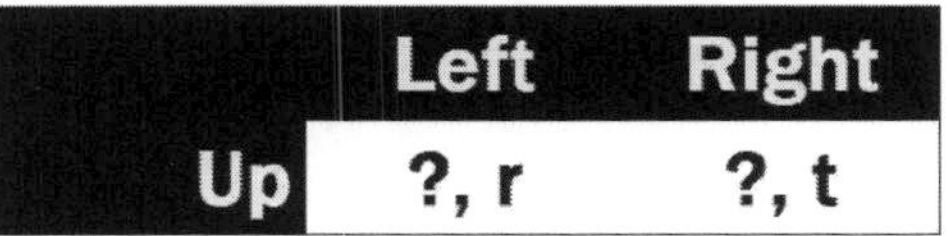

	Left	Right
Up	?, r	?, t

Player 2 should select right, as t beats r.

Alternatively, consider player 2's best response to down:

	Left	Right
Down	?, s	?, p

Once more, right is a better choice for player 2, as p is greater than s. Thus, right strictly dominates left.

From chapter 1, we know no mixed strategy Nash equilibrium exists. Players cannot use strictly dominated strategies in equilibrium. As such, the players cannot choose up or left in the generalized prisoner's dilemma. That leaves a single playable strategy for both players, meaning they cannot mix in equilibrium.

We can also show no mixed strategy Nash equilibrium exists by demonstrating that the indifference conditions from the mixed strategy algorithm produce invalid mixed strategies. When solving games with variables instead of numbers, it is easy to overlook instances of strict dominance. Fortunately, when we run the mixed strategy algorithm, we will eventually encounter a contradiction. Thus, even if we overlook strict dominance the first time, the algorithm will eventually save us.

To see this in practice, consider player 2's expected utility for playing left as a function of player 1's mixed strategy σ_{up}:

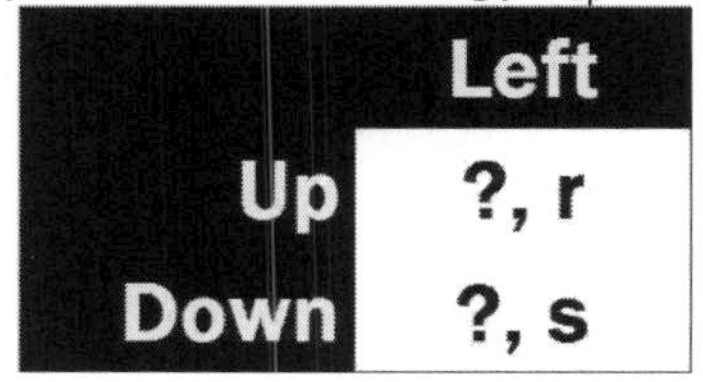

	Left
Up	?, r
Down	?, s

Player 2 earns r with probability σ_{up} and s with probability $1 - \sigma_{up}$. As an equation:

$EU_{left} = (r)(\sigma_{up}) + (s)(1 - \sigma_{up})$

Let's switch to right:

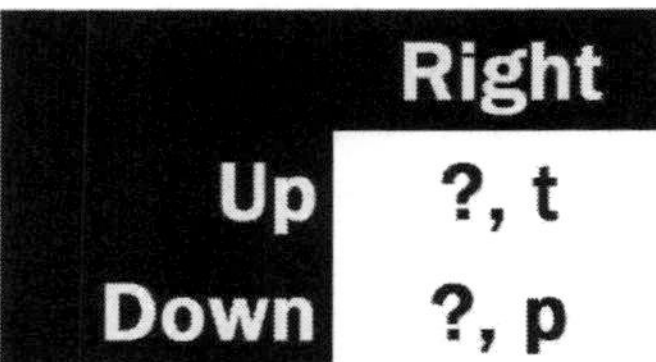

If player 2 selects right as a pure strategy, she earns t with probability σ_{up} and p with probability $1 - \sigma_{up}$. As an equation:

$EU_{right} = (t)(\sigma_{up}) + (p)(1 - \sigma_{up})$

To find player 1's mixed strategy that leaves player 2 indifferent between her pure strategies, we set player 2's expected utility for left equal to her expected utility for right and solve for σ_{up}:

$EU_{left} = EU_{right}$
$EU_{left} = (r)(\sigma_{up}) + (s)(1 - \sigma_{up})$
$EU_{right} = (t)(\sigma_{up}) + (p)(1 - \sigma_{up})$
$(r)(\sigma_{up}) + (s)(1 - \sigma_{up}) = (t)(\sigma_{up}) + (p)(1 - \sigma_{up})$
$r\sigma_{up} + s - s\sigma_{up} = t\sigma_{up} + p - p\sigma_{up}$
$r\sigma_{up} + p\sigma_{up} - s\sigma_{up} - t\sigma_{up} = p - s$
$\sigma_{up}(r + p - s - t) = p - s$

Before we solve for σ_{up}, we must check whether $r + p - s - t$ can equal zero. We can rewrite it as $(p - s) + (r - t)$. Since $p > s$, $p - s$ is positive. Similarly, since $t > r$, $r - t$ is negative. When we sum a positive number and a negative number, three things are possible: the result is positive, the result is negative, or the result is zero. All three of these cases can occur in practice depending on the size of each of the numbers. This leaves us with three separate cases to check.

<u>$r + p - s - t = 0$</u>

Let's start with the tricky one. When $r + p - s - t = 0$, we cannot divide by it to solve for σ_{up}. However, we can substitute 0 into the equation to see that there is going to be a problem:

$\sigma_{up}(r + p - s - t) = p - s$
$r + p - s - t = 0$
$\sigma_{up}(0) = p - s$
$0 = p - s$
$s = p$

Recall that p is greater than s. But the mixed strategy algorithm told us that s is equal to p. Nonsense! This is our algorithm's way of telling us that no probability distribution can make player 2 indifferent between her two pure strategies. Thus, when $r + p - s - t = 0$, no MSNE exists.

$\underline{r + p - s - t < 0}$

If $r + p - s - t$ is less than zero, we can divide by it. Let's do just that:

$\sigma_{up}(r + p - s - t) = p - s$
$\sigma_{up} = (p - s)/(r + p - s - t)$

Now we have to check whether $(p - s)/(r + p - s - t)$ can be part of a valid probability distribution. If so, the numerator and denominator must both be positive or both be negative. We know $p - s$ is positive, so $r + p - s - t$ must be positive as well. But we just assumed that $r + p - s - t < 0$, so the denominator is negative. When we divide a positive number by a negative number, we end up with a negative number. Probabilities cannot be negative. Thus, no mixed strategy will work in this case.

$\underline{r + p - s - t > 0}$

Finally, suppose $r + p - s - t$ is positive. We can therefore divide by it just as we did before:

$\sigma_{up}(r + p - s - t) = p - s$
$\sigma_{up} = (p - s)/(r + p - s - t)$

We know the numerator and denominator are both positive. The final step is to check whether the numerator is less than the denominator:

$p - s < r + p - s - t$
$0 < r - t$
$t < r$

However, since $t > r$, the numerator is greater than the denominator. Thus, $(p - s)/(r + p - s - t)$ is greater than 1 and in turn is not a valid probability. So this case fails as well.

Combining all three of these cases together, we know that regardless of the specific values of the exogenous variables, player 1 cannot induce player 2 to be indifferent between her pure strategies. Specifically, she will always prefer to play right, since right strictly dominates left.

We could run through the same process for player 2's mixed strategy, and we would find the same thing; after all, the game is symmetric. Therefore, no player can mix in this game, leaving <down, left> as the unique Nash equilibrium.

3.2.3: Generalized Deadlock

Let's reinforce these same principles with deadlock. Recycling the payoffs from the prisoner's dilemma, let $T > R > P > S$ and $t > r > p > s$. If we switch each player's temptation payoff for the sucker's payoff, deadlock results:

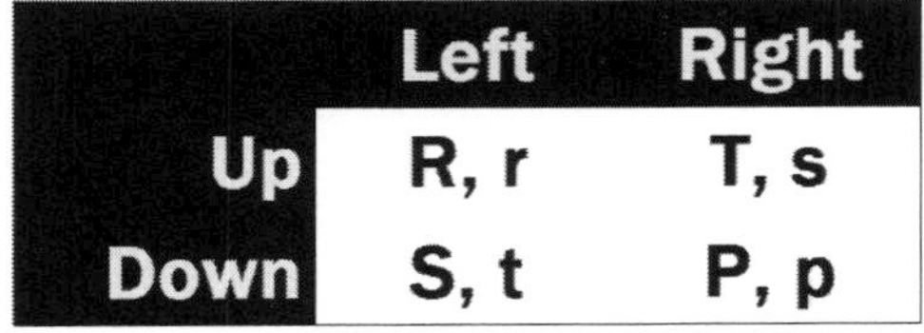

	Left	Right
Up	R, r	T, s
Down	S, t	P, p

This time around, up strictly dominates down and left strictly dominates right. Let's confirm player 2's strict dominance, starting with her response to up:

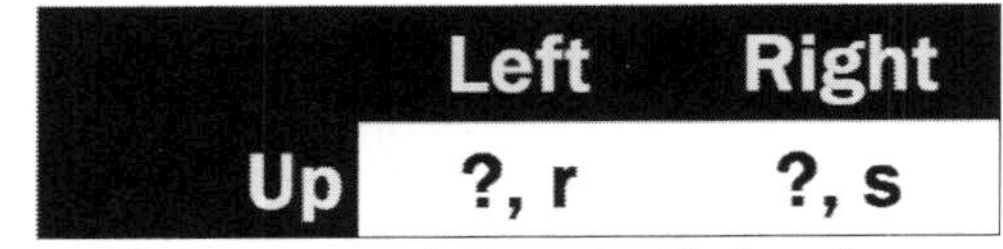

	Left	Right
Up	?, r	?, s

Player 2 earns r if she moves left and s if she chooses right. Since $r > s$, left is her best response.

Now consider down:

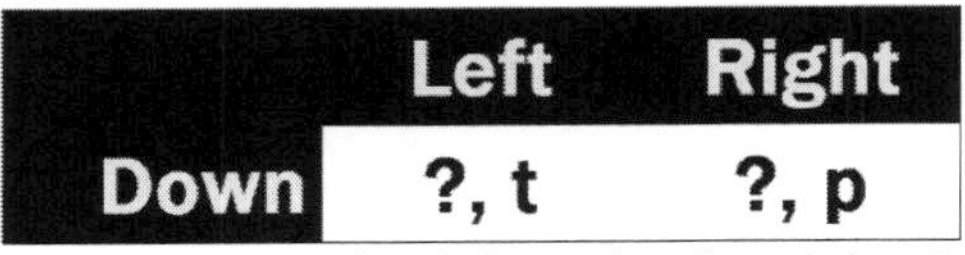

	Left	Right
Down	?, t	?, p

This time, player 2 earns t for left and p for right. Because $t > p$, her best response is left. Thus, she ought to play left regardless of what player 1 selects.

Player 1 has a similar dominant strategy. Suppose player 1 selected left:

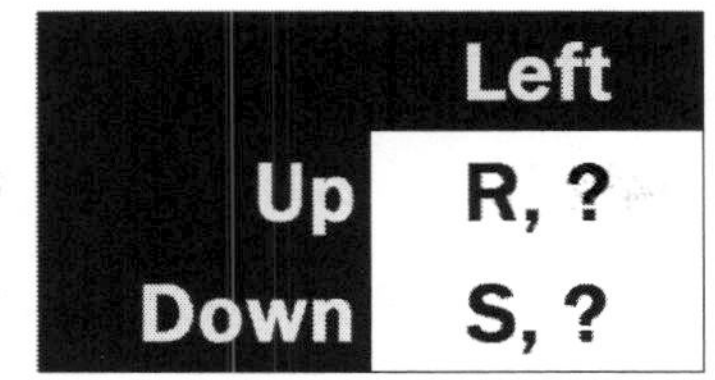

	Left
Up	R, ?
Down	S, ?

Since R > S, player 1 plays up in response to left.

Now consider player 1's options if player 2 chooses right:

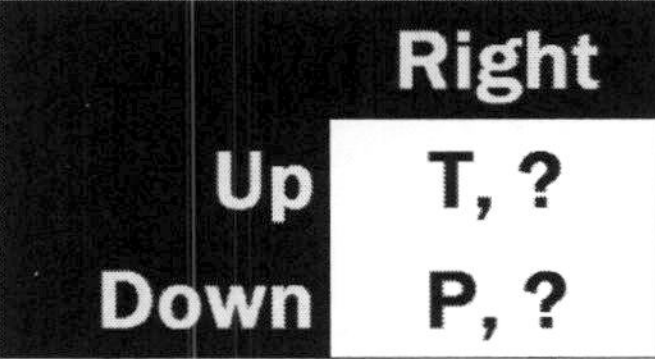

	Right
Up	T, ?
Down	P, ?

He still prefers up, as T > P.

Compiling these responses, we know player 1 should select up regardless of player 2's strategy. In other words, up strictly dominates down. Since left strictly dominates right for player 2, <up, left> is the unique Nash equilibrium of the game.

But even if we missed the strict dominance solution, the mixed strategy algorithm will come to our rescue. To see this, we can calculate player 1's expected utility as a function of player 2's mixed strategy σ_{left}.

To start, suppose player 1 chose up:

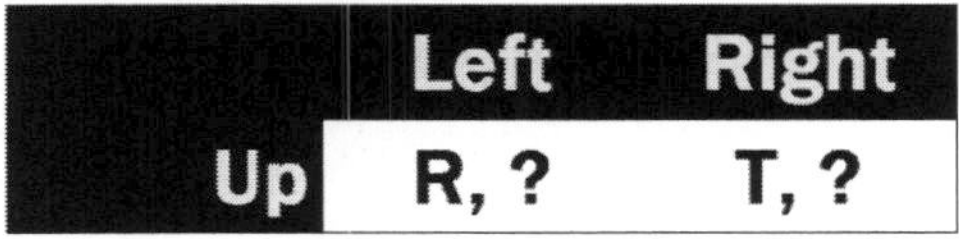

	Left	Right
Up	R, ?	T, ?

Here, he earns R with probability σ_{left} and T with probability $1 - \sigma_{left}$. As an equation:

$$EU_{up} = (\sigma_{left})(R) + (1 - \sigma_{left})(T)$$

And if player 1 selected down instead:

	Left	Right
Down	S, ?	P, ?

This time, he earns S with probability σ_{left} and P with probability $1 - \sigma_{left}$. As an equation:

$$EU_{down} = (\sigma_{left})(S) + (1 - \sigma_{left})(P)$$

Then we set player 1's expected utility for up equal to his expected utility for down and solve for σ_{left}:

$EU_{up} = EU_{down}$
$EU_{up} = (\sigma_{left})(R) + (1 - \sigma_{left})(T)$
$EU_{down} = (\sigma_{left})(S) + (1 - \sigma_{left})(P)$
$(\sigma_{left})(R) + (1 - \sigma_{left})(T) = (\sigma_{left})(S) + (1 - \sigma_{left})(P)$
$R\sigma_{left} + T - T\sigma_{left} = S\sigma_{left} + P - P\sigma_{left}$
$R\sigma_{left} + P\sigma_{left} - T\sigma_{left} - S\sigma_{left} = P - T$
$\sigma_{left}(R + P - T - S) = P - T$

Before we divide by $R + P - T - S$, we must check whether $R + P - T - S$ is equal to zero. Unfortunately, we fall into the same indeterminacy that we saw in the prisoner's dilemma. $P - T$ is negative, but $R - S$ is positive. Summing those together sometimes creates a positive number, sometimes creates a negative number, and sometimes results in 0. We therefore need to separate these possibilities into three cases.

$R + P - T - S = 0$

Here, we cannot divide to solve for σ_{left}. However, we can substitute $R + P - T - S = 0$ into the equation:

$\sigma_{left}(R + P - T - S) = P - T$
$R + P - T - S = 0$
$\sigma_{left}(0) = P - T$
$0 = P - T$
$T = P$

But $T > P$, so we have derived a contradiction. Thus, if $R + P - T - S = 0$, no mixture satisfies the mixed strategy algorithm, and we can switch to the next case:

$R + P - T - S > 0$

This time, we can divide by $R + P - T - S$:

$\sigma_{left}(R + P - T - S) = P - T$
$\sigma_{left} = (P - T)/(R + P - T - S)$

Since $P < T$, we know $P - T$ is negative. To be a valid probability, the denominator must therefore also be negative. However, we assumed it was positive at the start of this case, so we have found a contradiction and can advance to the final case:

<u>$R + P - T - S < 0$</u>

Now when we divide $P - T$ by $R + P - T - S$, the result is positive. The last thing to check is whether the magnitude of the numerator is less than the magnitude of the denominator. Since both the numerator and denominator are negative here, this means $P - T$ must be *greater* than $R + P - T - S$:

$P - T > R + P - T - S$
$0 > R - S$
$S > R$

However, R is greater than S by definition, so the probability distribution fails once again.

Combining all three cases together, no mixed strategy satisfies the indifference condition. Therefore, a general form of deadlock has no mixed strategy Nash equilibrium.

Takeaway Points

1) MSNE are sensitive to slight changes in payoffs. Using generalized games with exogenous variables allows us to solve all versions of a game with one calculation.
2) We must be careful to show that probabilities the mixed strategy algorithm produces are valid.

Lesson 3.3: Knife-Edge Equilibria

When we use game theory to model real world phenomenon, we want to use payoffs that actually represent the real world preferences of the players. After all, payoffs affect a game's equilibria. If we use bizarre payoffs, we will find bizarre equilibria. In turn, our game will falsely predict the behavior of the actors in the model.

Using exogenous variables in payoffs sometimes leads to cases where an equilibrium exists for only a single configuration of the payoffs; increasing or decreasing a single payoff by a tiny amount makes that equilibrium completely disappear. We refer to such an equilibrium as a *knife-edge equilibrium*, as they precariously rest on the skinny edge of a single number. We generally believe these equilibria are unlikely to occur in the real world since they are only possible if the numbers magically align themselves in a perfect way. Thus, we generally ignore knife-edge equilibria.

This lesson shows how to spot knife-edge equilibria and further explains why we intentionally toss them aside.

3.3.1: The Hawk-Dove Game

Two animals are in conflict over some good worth v > 0. Simultaneously, they choose whether to behave like hawks or doves. Hawks are willing to fight over the good. Doves are not. So if one animal chooses hawk and the other selects dove, the hawk takes the entire good and the dove receives nothing. If both act like hawks, they fight a battle over the good. Each wins the good with probability 1/2 but both pay a cost of conflict c > 0. Finally, if both play dove, they split the good evenly.

Let's look at the matrix:

	Hawk	Dove
Hawk	v/2 - c, v/2 - c	v, 0
Dove	0, v	v/2, v/2

In the past couple of lessons, the games have had stable equilibria even as the exogenous variables changed. For example, in the prisoner's dilemma, regardless of the particular values for T, R, P, and S, both players always played their strictly dominant strategies in the Nash equilibrium. Likewise, in the generalized battle of the sexes game, two PSNE and one MSNE always existed regardless of the values of A, B, and C. Although the exact mixtures of the MSNE changed as a function of A, B, and C, the players could always mix in equilibrium.

The hawk-dove game is not as straightforward. When the value of v is high, the players have a strictly dominant strategy and always select that one. But when the value of v is low, the game has two PSNE and one MSNE.

Let's see why changing the value of v leads to different equilibria. First, consider player 1's best response to dove:

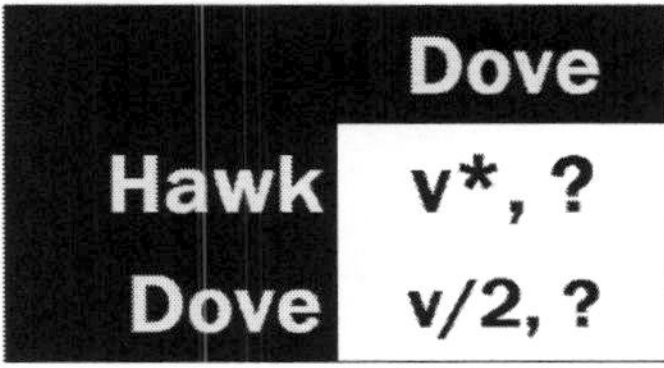

	Dove
Hawk	v*, ?
Dove	v/2, ?

Since v is greater than 0, v is always greater than half of v. As such, player 1's best response to dove is hawk regardless of the specific value of v.

The same is true for player 2's best response to dove:

	Hawk	Dove
Dove	?, v*	?, v/2

Again, v is always greater than half of v, so hawk is player 2's best response.

However, each player's best response to hawk depends on the value of v. Observe player 1's decision:

	Hawk
Hawk	v/2 - c, ?
Dove	0, ?

When v/2 is greater than c, hawk is player 1's best response. But when v/2 is less than c, dove is his best response. Life gets complicated when v/2 exactly equals c, as *both* hawk and dove are best responses to hawk in that instance.

The same, of course, applies to player 2's best responses to player 1 playing hawk:

	Hawk	Dove
Hawk	?, v/2 - c	?, 0

Given such a dynamic, the best course of action is to split the game into three cases: v/2 > c, v/2 < c, and v/2 = c.

Let's start with v/2 > c. Here, hawk is each player's best response to the other playing hawk. But hawk is also each player's best response to dove. Therefore, hawk strictly dominates dove for v/2 > c. We know both players must pick hawk in equilibrium. In essence, the game turns into a prisoner's dilemma, in which "hawk" is "confess" and "dove" is "keep quiet." In equilibrium, both players earn v/2 – c. However, if they could credibly commit to playing dove, both players improve their payoffs to v/2. Unfortunately, for the same reason as in the prisoner's dilemma, the <dove, dove> outcome is inherently unstable; both players have incentive to deviate to hawk.

Now consider v/2 < c. Dove is now each player's best response to hawk. Consequently, with best responses marked, the game looks like this:

	Hawk	Dove
Hawk	v/2 - c, v/2 - c	v*, 0*
Dove	0*, v*	v/2, v/2

Thus, two pure strategy Nash equilibria exist: <hawk, dove> and <dove, hawk>. But since no player has a single strictly dominant strategy, we must run the mixed strategy algorithm. On the bright side, the symmetry of the game means that we only need to solve for one player's mixed strategy; the other player's will be the same.

Let's begin by finding player 1's expected utility for hawk:

	Hawk	Dove
Hawk	v/2 - c, ?	v, ?

Player 1 earns v/2 – c with probability σ_{hawk} and v with probability 1 – σ_{hawk}. As an equation:

$$EU_{hawk} = (\sigma_{hawk})(v/2 - c) + (1 - \sigma_{hawk})(v)$$

Now consider player 1's possible payoffs if he selects dove:

	Hawk	Dove
Dove	0, ?	v/2, ?

This time, he earns 0 with probability σ_{hawk} and v/2 with probability 1 – σ_{hawk}. As an equation:

$$EU_{dove} = (\sigma_{hawk})(0) + (1 - \sigma_{hawk})(v/2)$$

To solve for player 2's equilibrium mixed strategy, we set those two expected utilities equal to one another and solve for σ_{hawk}:

$$EU_{hawk} = EU_{dove}$$
$$EU_{hawk} = (\sigma_{hawk})(v/2 - c) + (1 - \sigma_{hawk})(v)$$

$EU_{dove} = (\sigma_{hawk})(0) + (1 - \sigma_{hawk})(v/2)$
$(\sigma_{hawk})(v/2 - c) + (1 - \sigma_{hawk})(v) = (\sigma_{hawk})(0) + (1 - \sigma_{hawk})(v/2)$
$(\sigma_{hawk})(v/2 - c) + (1 - \sigma_{hawk})(v) = (1 - \sigma_{hawk})(v/2)$
$v\sigma_{hawk}/2 - c\sigma_{hawk} + v - v\sigma_{hawk} = v/2 - v\sigma_{hawk}/2$
$-v\sigma_{hawk}/2 - c\sigma_{hawk} + v = v/2 - v\sigma_{hawk}/2$
$-c\sigma_{hawk} + v = v/2$
$c\sigma_{hawk} = v/2$
$\sigma_{hawk} = v/2c$

Let's verify that v/2c is a valid probability distribution. The c and v are both positive, which implies 2c is positive. Thus, the numerator and denominator are both positive. That fulfills the first requirement. Second, we must confirm that the numerator is less than the denominator, or $v < 2c$. Recall that for this particular case, we have already assumed that $v/2 < c$. If we multiply both sides by 2, we immediately arrive at our required condition: $v < 2c$. As such, we have a valid probability distribution.

Thus, if a player chooses hawk with probability v/2c, the other player is indifferent between both of his or her strategies. In turn, in the case where $v < 2c$, a MSNE exists in which both players select hawk with probability v/2c and choose dove with complementary probability.

Finally, we must investigate the case where $v/2 = c$. Recall that each player's payoff for the <hawk, hawk> outcome was $v/2 - c$. Thus, when we substitute $v/2 = c$, that payoff reduces to 0.

Making that change, here is the matrix for $v/2 = c$:

	Hawk	Dove
Hawk	0, 0	v, 0
Dove	0, v	v/2, v/2

As we saw earlier, each player now is indifferent between hawk and dove if the other player selects hawk as a pure strategy. Consequently, this version of the hawk-dove game has three PSNE: <hawk, hawk>, <hawk, dove>, and <dove, hawk>. We can observe this with the best responses marked:

	Hawk	Dove
Hawk	0*, 0*	v*, 0*
Dove	0*, v*	v/2, v/2

Unfortunately, the indifference between hawk and dove when the other player selects hawk leads to infinitely many equilibria in partially mixed strategies. To see this, suppose player 1 mixes. That is, he plays hawk with probability σ_{hawk} and dove with probability $1 - \sigma_{hawk}$, where $0 < \sigma_{hawk} < 1$. (This restriction on σ_{hawk} requires player 1 to actually mix and not play hawk or dove as a pure strategy.) If player 2 plays hawk as a pure strategy, player 1's mixture is a best response, since he earns 0 regardless of his choice.

As such, we only need to see if player 2's hawk strategy is a best response to player 1's mixture. Let's begin by calculating her expected utility for hawk:

	Hawk
Hawk	?, 0
Dove	?, v

Player 2 earns 0 with probability σ_{hawk} and v with probability $1 - \sigma_{hawk}$. As an equation:

$$EU_{hawk} = (\sigma_{hawk})(0) + (1 - \sigma_{hawk})(v)$$

Now consider player 2's payoffs if she chooses dove:

	Dove
Hawk	v, ?
Dove	v/2, ?

This time, she earns 0 with probability σ_{hawk} and v/2 with probability σ_{hawk}. As an equation:

$$EU_{dove} = (\sigma_{hawk})(0) + (1 - \sigma_{hawk})(v/2)$$

As such, player 2 is willing to play hawk if her expected utility for hawk is at least as great as her expected utility for dove. Some algebra shows this is true:

$$EU_{hawk} \geq EU_{dove}$$
$$EU_{hawk} = (\sigma_{hawk})(0) + (1 - \sigma_{hawk})(v)$$
$$EU_{dove} = (\sigma_{hawk})(0) + (1 - \sigma_{hawk})(v/2)$$
$$(\sigma_{hawk})(0) + (1 - \sigma_{hawk})(v) \geq (\sigma_{hawk})(0) + (1 - \sigma_{hawk})(v/2)$$

$(1 - \sigma_{hawk})(v) \geq (1 - \sigma_{hawk})(v/2)$
$v - v\sigma_{hawk} \geq v/2 - v\sigma_{hawk}/2$
$v/2 \geq v\sigma_{hawk}/2$
$\sigma_{hawk} \leq 1$

In fact, σ_{hawk} must be less than 1, so this inequality strictly holds. Player 2's hawk strategy is the unique best response to any mixture of player 1's. In turn, we have infinitely many equilibria in which player 1 mixes between hawk and dove and player 2 selects hawk as a pure strategy. But since this game is symmetric, player 2 can also mix between hawk and dove in any fashion while player 1 chooses hawk as a pure strategy in equilibrium.

However, the partially mixed strategy Nash equilibria rely on the knife-edge condition of v/2 = c. If v/2 is even slightly greater than or slightly less than c, these equilibria completely disappear.

3.3.2: Why Are Knife-Edge Equilibria Unrealistic?

Most of the time, game theorists model situations that are naturally occurring and do not have laboratory controls. In the hawk-dove game, we thought of two animals fighting for control over a good. What are the odds that their cost of fighting was exactly v/2? It would be a minor miracle if the stars aligned in such a way to produce that result.

Going a step further, if we are explicit with our assumptions, the probability such a case exists equals *exactly* zero. That is, such a case *cannot* occur naturally.

Our theoretical justification comes from probability theory. Think of the universe of situations that the hawk-dove game represents. We can think of each individual situation's cost parameter c as coming from a continuous probability distribution, where c must be greater than 0 at minimum and has some upper cap C. Then the probability of observing a cost parameter c between any two values a and b (where a < b) is the integral of the probability density function between a and b.

Mathematically, let f(x) be that probability density function. Then the probability c falls between a and b is as follows:

$$Pr[a \leq c \leq b] = \int_a^b f(x)dx$$

The probability density function tells us that the probability c equals exactly v/2 is the integral of f(x) from v/2 to v/2:

$$Pr[v/2 \leq c \leq v/2] = \int_{v/2}^{v/2} f(x)dx$$

However, the rules of integration tell us that the integral from a single value to that same value equals zero. Essentially, we are looking for the area under the probability density function curve at a single point. But that area has no width and hence has no area. Thus, the probability of observing a c exactly equal to v/2 is zero.

Moving one step further, if the probability of something occurring is zero, then it never happens. If something never happens, there is no reason to study it. As such, when we encounter knife-edge conditions such as v/2 = c in the hawk-dove game, we tend to ignore them. After all, we cannot rest our theories on such a non-existent case, so analyzing these knife-edge conditions is a waste of effort and paper.

Although the next section will address a reason why we might care about knife-edge conditions, let it be known that this norm against researching knife-edge equilibria is a good thing. Knife-edge conditions induce indifference, which often leads to instances of weak dominance. Observe that hawk weakly dominates dove if v/2 equals c:

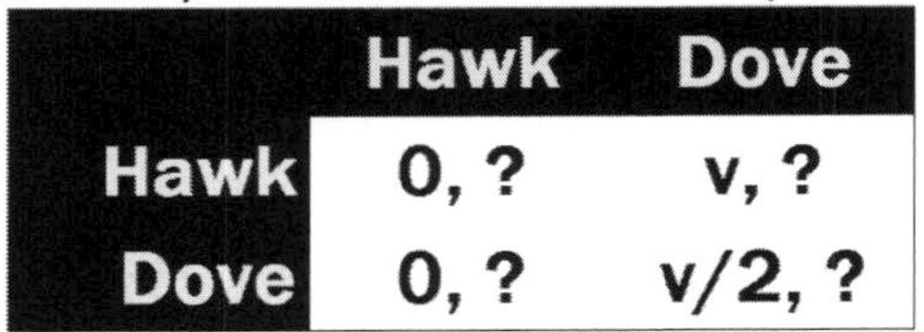

	Hawk	Dove
Hawk	0, ?	v, ?
Dove	0, ?	v/2, ?

If player 2 chooses hawk, hawk and dove give the same payoff to player 1. However, if player 2 selects dove, hawk beats dove for player 1. Thus, hawk always provides at least as great of a payoff as dove for player 1 and occasionally provides more. By definition, hawk weakly dominates dove here. The same is true for player 2.

As chapter 1 explained at length, weak dominance is the bane of our existence. We want to avoid solving these games whenever possible. Knife-edge conditions provide a theoretically justifiable reason to ignore such situations.

3.3.3: When Knife-Edge Conditions Are Important

We can safely skip knife-edge conditions of games that occur naturally. However, game masters sometimes fabricate the rules. For example, recall back to the take-or-share game from Lesson 1.8. Many game shows require their contestants to simultaneously decide to share a good or attempt to steal all of it for themselves. If we let the value of the good be v and we assume the players only want to maximize their share of the good, the generalized form of the game looks as follows:

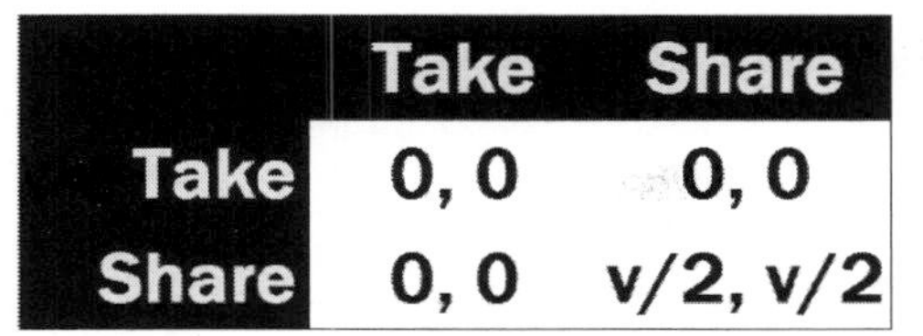

	Take	Share
Take	0, 0	0, 0
Share	0, 0	v/2, v/2

This is identical to the hawk-dove game when $v/2 = c$, except the word "take" appears in place of "hawk" and "dove" has become "share." Unfortunately, we cannot invoke the knife-edge exception here. A person (the game show creator) fabricated an environment that constrained $v/2$ to be equal to c. However, absent a scenario where someone is actually in control of the game in this manner, we can ignore knife-edge equilibria.

One more caveat to stress: we can only make claims about knife-edge conditions when referring to *exogenous* variables. We have seen and will see many games where an equilibrium rests on a knife-edge strategy *endogenously* selected by the players. For example, recall the matching pennies game from Lesson 1.5. In that game's unique Nash equilibrium, the players must play heads exactly half the time and tails exactly half the time. Any extra weight breaks the indifference condition, and we lose the equilibrium.

Even so, this is theoretically justifiable. The players actively choose these strategies in response to the constraints of their environment. In contrast, we have no theoretical justification for exogenous variables creating a knife-edge condition. Humans can make strategic decisions; nature cannot.

Takeaway Points

1) A *knife-edge equilibrium* is an equilibrium that occurs for an exact set of payoffs. Slightly increasing or decreasing a single payoff completely eliminates the equilibrium.
2) Knife-edge conditions do not occur naturally, so we usually skip solving them.
3) If we were to solve them, they usually take a while—knife-edge conditions lead to weakly dominated strategies.

Lesson 3.4: Comparative Statics

At its core, game theory is the study of altering the strategic dimensions of an environment. We want to know how subtle changes to a game affect how players behave. For example, during soccer penalty kick, the striker can aim to the left side of the goal or the right side of the goal. Suppose he is extremely accurate on the right side but has problems with his left side. Aware of his weakness, the striker has spent the past two months practicing on the left side. How should this affect his strategy?

Although such scenarios seem straightforward at the outset, game theory has a notorious history of finding optimal but counterintuitive strategies. (As we will see shortly, this penalty kicks example is one of them.) We began introducing the possibility of a fluid environment in this chapter by adding exogenous variables to payoff matrices. In this lesson, we begin analyzing the actual change.

The study of such changes is called *comparative statics*. In essence, we take one environment, make a slight tweak to it, and compare the outcomes of those two games. Using this method, we can discover how manipulating games affects a player's outcome or the welfare of a society.

Fortunately, calculating comparative statics is an easy (if time consuming) process. Unfortunately, we now enter a new level of computationally intensive game theory, as we must take derivatives to calculate these comparative statics.

Our method of calculating comparative statics is as follows:

1) Solve for the game's equilibria.
2) Calculate the element of interest. (This could be the probability a player selects a particular strategy, the probability the players reach a certain outcome, or a player's expected utility.)
3) Take the derivative of that element of interest with respect to the exogenous variable we want to manipulate.
4) Use that derivative to see how changing the exogenous variable affects the element of interest.

The process may seem difficult to comprehend without some examples. Luckily, this lesson is full of them. Calculating comparative statics will seem easy by the end of this lesson.

3.4.1: Penalty Kicks

Consider the strategic interaction of soccer penalty kicks. The kicker can choose whether to aim left or right. Since the shot moves so quickly, the goalie effectively chooses simultaneously whether to dive left or right to stop the ball. The goalie wants to guess correctly; the kicker wants to make the goalie guess incorrectly.

Let's begin our analysis with a superhuman kicker and a superhuman goalie. The kicker has perfect accuracy; if the goalie guesses incorrectly, the kicker always buries the ball into the net. However, if the goalie guesses correctly, he always stops the shot. Then we can use the following matrix to analyze the game:

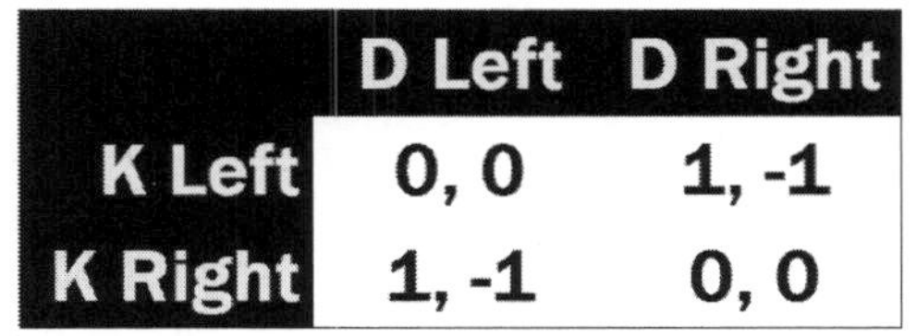

	D Left	D Right
K Left	0, 0	1, -1
K Right	1, -1	0, 0

Although the payoffs are different, the general structure of this game is identical to the matching pennies framework presented in Lesson 1.5. The kicker wants to mismatch directions; the goalie wants to match directions. Thus, no PSNE exists.

And just like matching pennies, the MSNE calls for both players to flip a coin. This clearly induces indifference on both sides. If the goalie dives left and dives right with equal frequency, then the kicker scores at the same rate regardless of which direction he aims. Likewise, if the kicker aims left half of the time and aims right half of the time, then the probability the goalie stops the shot is the same regardless if she dives left or right. Thus, neither player has a profitable deviation, which therefore makes this a Nash equilibrium.

But what if the kicker had a weak side? Perhaps if the kicker aims left, he will miss some percentage of the time. Thus, even if the goalie guesses incorrectly, the kicker might not score. Let the kicker's accuracy on his weak left side be x, where $0 < x < 1$. Then we can use the following matrix to represent the interaction:

	D Left	D Right
K Left	0, 0	x, -x
K Right	1, -1	0, 0

Before we solve this game, make a prediction. *As the kicker's accuracy on his left side increases, will he kick to that side more frequently?* This question is obviously important to a soccer player, but only comparative statics can prove the correct answer.

Let's begin the comparative statics process. Recall that the first step to deriving comparative statics is to find a game's equilibria. We ought to begin here by marking best responses:

	D Left	D Right
K Left	0, 0*	x*, -x
K Right	1*, -1	0, 0*

Despite the additions of x and –x to the game, the best responses have the same pattern as in matching pennies. The goalie still wants to match the kicker's behavior, while the kicker still wants to play the opposite of the goalie's strategy. Therefore, the game lacks a PSNE. In turn, we must look for its MSNE.

Let's start by finding the kicker's expected utility for aiming left:

	D Left	D Right
K Left	0, ?	x, ?

If the goalie dives left, he earns 0. If she dives right, he earns x. As an equation:

$EU_{KL} = (\sigma_{DL})(0) + (1 - \sigma_{DL})(x)$

Now we move to the kicker's expected utility for kicking right:

	D Left	D Right
K Right	1, ?	0, ?

This time, he earns 1 if the goalie dives left and 0 if she dives right. As an equation:

$EU_{KR} = (\sigma_{DL})(1) + (1 - \sigma_{DL})(0)$

The last step is to set these two expected utilities equal to each other and solve for σ_{DL}.

$EU_{KL} = EU_{KR}$
$EU_{KL} = (\sigma_{DL})(0) + (1 - \sigma_{DL})(x)$
$EU_{KR} = (\sigma_{DL})(1) + (1 - \sigma_{DL})(0)$
$(\sigma_{DL})(0) + (1 - \sigma_{DL})(x) = (\sigma_{DL})(1) + (1 - \sigma_{DL})(0)$
$(1 - \sigma_{DL})(x) = (\sigma_{DL})(1)$

$x - x\sigma_{DL} = \sigma_{DL}$
$\sigma_{DL} + x\sigma_{DL} = x$
$\sigma_{DL}(1 + x) = x$
$\sigma_{DL} = x/(1 + x)$

Because x is a positive number, 1 + x is also a positive number. 1 + x is also greater than x, so this is a valid probability distribution.

Knowing this, we can safely move to the goalie's mixed strategy. Let's start with her expected utility for diving left:

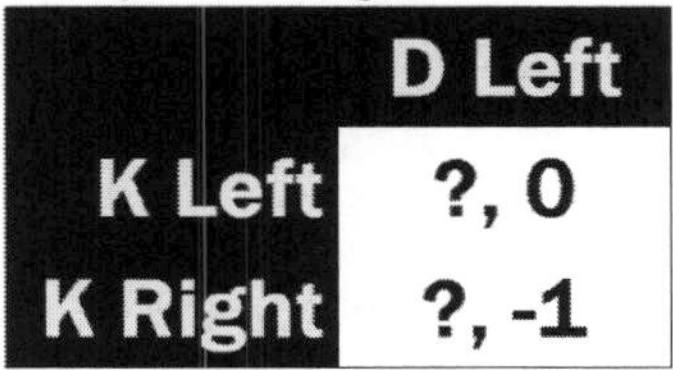

	D Left
K Left	?, 0
K Right	?, -1

The goalie makes the save and earns 0 with probability σ_{KL}. With probability $1 - \sigma_{KL}$, she guesses incorrectly and earns -1. As an equation:

$EU_{DL} = (\sigma_{KL})(0) + (1 - \sigma_{KL})(-1)$

Let's switch to diving right:

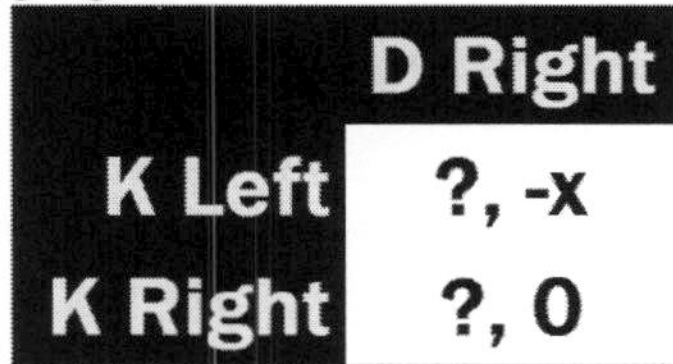

	D Right
K Left	?, -x
K Right	?, 0

This time, the goalie earns -x with probability σ_{KL} and 0 with probability $(1 - \sigma_{KL})$. As an equation:

$EU_{DR} = (\sigma_{KL})(-x) + (1 - \sigma_{KL})(0)$

Just like before, we set these expected utilities equal to each other and solve for σ_{KL}:

$EU_{DL} = EU_{DR}$
$EU_{DL} = (\sigma_{KL})(0) + (1 - \sigma_{KL})(-1)$
$EU_{DR} = (\sigma_{KL})(-x) + (1 - \sigma_{KL})(0)$
$(\sigma_{KL})(0) + (1 - \sigma_{KL})(-1) = (\sigma_{KL})(-x) + (1 - \sigma_{KL})(0)$
$(1 - \sigma_{KL})(-1) = (\sigma_{KL})(-x)$
$-1 + \sigma_{KL} = -x\sigma_{KL}$
$\sigma_{KL} + x\sigma_{KL} = 1$

$\sigma_{KL}(1 + x) = 1$
$\sigma_{KL} = 1/(1 + x)$

This is a valid probability distribution: 1 and 1 + x are both positive and 1 + x is greater than 1.

Thus, we found the MSNE: the goalie dives left with probability x/(1 + x) while the kicker aims left with probability 1/(1 + x).

Recall the step-by-step process to calculate comparative statics. The first step is to calculate the equilibrium. We can check that part off. The second step is to calculate the element of interest. We want to know how the kicker's strategy changes as his aim toward the left side improves. To do that, we must know the probability he kicks left. Fortunately, the mixed strategy Nash equilibrium explicitly tells us that this probability is 1/(1 + x), so the first step took care of the second step as well.

The third step is to take the derivative. Since x represents the kicker's accuracy, we take the derivative of 1/(1 + x) with respect to x. This requires the quotient rule:

$f(x) = 1/(1 + x)$
$f'(x) = [(1)'(1 + x) - (1)(1 + x)']/(1 + x)^2$
$f'(x) = [(0)(1 + x) - (1)(1)]/(1 + x)^2$
$f'(x) = -1/(1 + x)^2$

Recall that x is bounded between 0 and 1. Thus, $-1/(1 + x)^2$ is always negative on that interval. Therefore, the probability the kicker aims left *decreases* as his accuracy to that side improves!

Most people guess the opposite. After all, why would improving your abilities on one side make you want to utilize that side less frequently?

The critical insight is that the kicker must factor in the goalie's strategic interaction. In a world where the goalie knows nothing about the kicker's weakness, the kicker should aim toward the stronger side. However, in this version of the game, the goalie is fully aware of the kicker's weakness, and she exploits that weakness by guarding the kicker's strong side more frequently. In turn, the kicker sees value in aiming toward his weak side: although he will often miss, the goalie will not be there to stop a well-placed shot as frequently. But when the kicker's accuracy on his left improves, the goalie can no longer camp out on the right. That makes the kicker willing to aim to the right more often, which is what the comparative static tells us will happen.

At least in this context, the kicker is only as good as his weakest link.

3.4.2: The Volunteer's Dilemma

This section explores a common explanation for the story of Kitty Genovese, a woman who stabbed outside of her apartment but received surprisingly little help from onlookers.

The story goes as follows. Two neighbors share an apartment building. Late one night, a deranged man stabs a woman in the alleyway below the apartment. She shrieks, waking both of the neighbors. The neighbors look out their windows and see the woman bleeding on the ground. Simultaneously and without communicating with one another, the neighbors individually must decide whether to call the police. If no one calls, the woman will die. If a single neighbor picks up the phone, an ambulance will arrive in time to save her life.

Let's suppose the neighbors are nice people: they value the woman's life worth 1 and her death worth 0. However, calling is costly to them, since they will be up all night talking to the police officers and risk retribution from the murderer. To reflect this, anyone who calls pays a cost c, where $0 < c < 1$. We can imagine c to be very small, as the neighbors would be very sad if they let the woman die. However, they both prefer the other one make the phone call. Hence this is a *volunteer's dilemma*: each neighbor only wants to volunteer to call if he or she knows the other one will not. As such, without communication, it is unclear who should pick up the phone.

Here is the payoff matrix:

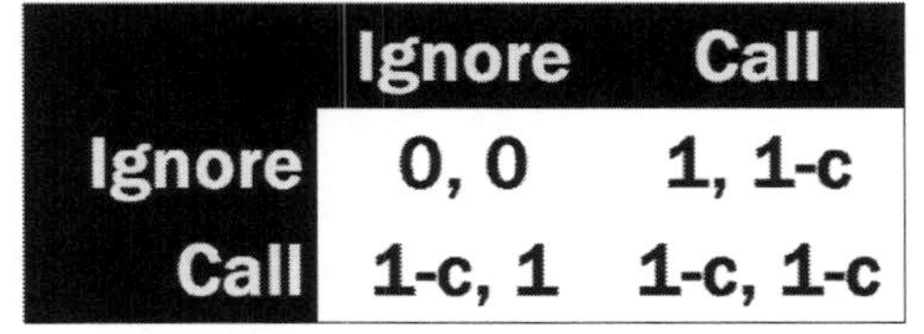

	Ignore	Call
Ignore	0, 0	1, 1-c
Call	1-c, 1	1-c, 1-c

Two PSNE exist: <call, ignore> and <ignore, call>. To see this, suppose the player calling in equilibrium switched to ignoring. Now no one calls, the woman dies, and both players earn 0. This is not a profitable deviation, as the caller previously earned 1 – c from calling, which is a positive amount. Likewise, the player ignoring in equilibrium would not want to change strategies. Help will already arrive thanks to the other player. If the ignorer switches to calling, he unnecessarily pays the cost.

Marking the best responses further illustrates these points:

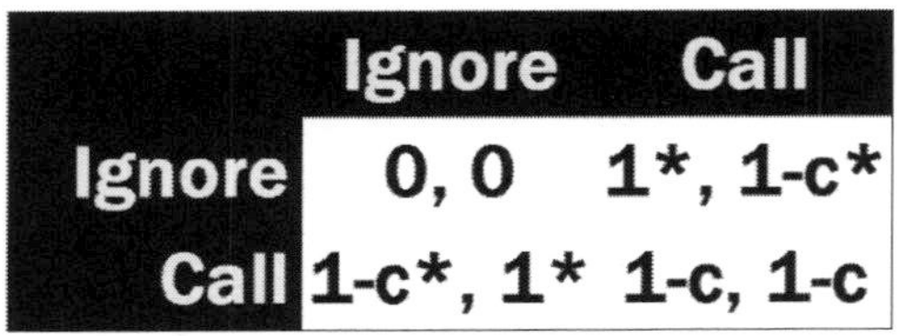

	Ignore	Call
Ignore	0, 0	1*, 1-c*
Call	1-c*, 1*	1-c, 1-c

Thus far, we have found two equilibria. However, Lesson 1.8 taught us that virtually all games have an odd number of equilibria, so we should look for a mixed strategy Nash equilibrium here as well. Let's start by finding player 1's expected utility for ignoring:

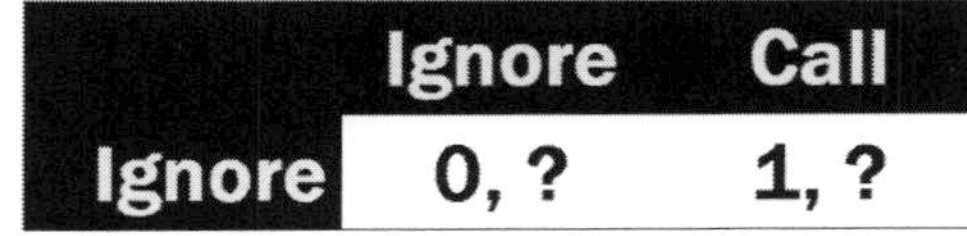

	Ignore	Call
Ignore	0, ?	1, ?

Let σ_{ignore} be the probability player 2 ignores. Then player 1 earns 0 with probability σ_{ignore} and 1 with probability $1 - \sigma_{ignore}$. As an equation:

$$EU_{ignore} = (\sigma_{ignore})(0) + (1 - \sigma_{ignore})(1)$$

Now consider the possibilities if player 1 calls:

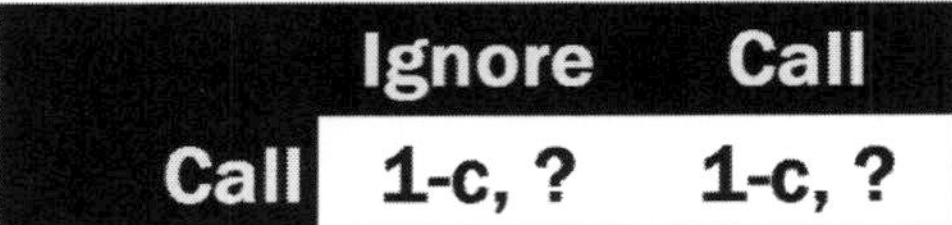

	Ignore	Call
Call	1-c, ?	1-c, ?

Interestingly, player 2's strategy is irrelevant here. If player 1 calls, he pays a cost and help arrives with certainty. Thus, he guarantees himself a payoff of $1 - c$:

$$EU_{call} = 1 - c$$

For our final step, we set these expected utilities equal to each other and solve for σ_{ignore}:

$$EU_{ignore} = EU_{call}$$
$$EU_{ignore} = (\sigma_{ignore})(0) + (1 - \sigma_{ignore})(1)$$
$$EU_{call} = 1 - c$$
$$(\sigma_{ignore})(0) + (1 - \sigma_{ignore})(1) = 1 - c$$
$$1 - \sigma_{ignore} = 1 - c$$
$$\sigma_{ignore} = c$$

Therefore, in the MSNE, player 1 ignores with probability c. But note that the game is symmetric. As such, player 2 adopts the exact same strategy. So both players ignore with probability c and call with probability $1 - c$.

This equilibrium gravely concerns the bleeding woman. If both players ignore with probability c, then *no one* calls with probability c^2. Thus, c^2 of the time, the woman dies in the alley even though both players would prefer to call if they knew this would be the eventual outcome.

Would the dying woman prefer selfish or benevolent witnesses? We can find out using comparative statics. Consider the cost parameter c to be a measure of selfishness. The higher its value, the more the witnesses value their own time compared to the woman's survival.

Since we already know that the probability no one calls and the woman dies equals c^2, we only need to take the derivative and analyze it. Fortunately, the power rule is easy to implement here:

$$f(c) = c^2$$
$$f'(c) = 2c$$

The derivative is 2c. Recall that c always takes on a positive value. Thus, the derivative is always increasing on the valid values. In turn, we can infer that the woman dies more frequently as c increases. We have reached a logical conclusion: the woman prefers selfless witnesses to selfish ones.

3.4.3: Comparative Statics of the Hawk-Dove Game

Recall that the hawk-dove game looked like this:

	Hawk	Dove
Hawk	v/2 - c, v/2 - c	v, 0
Dove	0, v	v/2, v/2

For $v/2 > c$, both players choose hawk in the unique equilibrium. When $v/2 < c$, three equilibria exist: <hawk, dove> and <dove, hawk> in pure strategies and a MSNE in which both select hawk with probability v/2c. Consider the following question:

Suppose the hawk-dove game is a model of crisis bargaining between two states, where the <hawk, hawk> outcome represents war. What can we say about the probability of war as a function of the cost of conflict?

Calculating comparative statics in the penalty kicks game and the volunteer's dilemma was relatively easy, as the equilibria only gradually changed as a function of the exogenous parameter. The hawk-dove game, however, has a drastic change in equilibria as c increases. When c is low, the states always choose hawk and thus end up in war with certainty. But once c crosses the v/2 threshold, the equilibria completely change; either one plays hawk and the other selects dove, or they mix. If they choose <hawk, dove> or <dove, hawk>, they avoid war entirely.

If they mix, their mixing probabilities depend on c, so we must calculate the appropriate derivative. Since they both select hawk with probability v/2c, the probability they fight a war equals (v/2c)(v/2c). To see how the mixed strategy Nash equilibrium evolves as a function of c, we must take the derivative of that probability:

$Pr(war) = (v/2c)(v/2c)$
$Pr(war) = v^2/4c^2$

Rather than suffer through the quotient rule, we can bring the c into the denominator by recalling that $1/x^n = x^{-n}$. Thus:

$Pr(war) = v^2/4c^2$
$Pr(war) = v^2c^{-2}/4$
$Pr(war)' = (-2)(v^2c^{-3})/4$
$Pr(war)' = -v^2c^{-3}/2$
$Pr(war)' = -v^2/2c^3$

Since both v and c are greater than 0, $v^2/2c^3$ is always positive, which means $-v^2/2c^3$ is always negative. Therefore, the probability of war is decreasing as a function of c.

To answer the original question, war is weakly decreasing as a function of c. Put differently, if we increase the cost of war, the expected probability of war will either remain the same or decrease. Two reasons prevent us from saying it is decreasing instead of weakly decreasing. First, if we increase c but maintain $c < v/2$, the players will continue playing <hawk, hawk> despite paying a larger cost. And second, if we increase c in the range $c > v/2$, the probability of war only decreases if the players are using the MSNE. If they are playing one of the PSNE, the probability of war remains 0.

Regardless, political scientists interested in interstate wars have created far more elaborate models of crisis bargaining and seen the same catch-22 result. Promoters of peace may want to limit the harm war inflicts on those unlucky enough to be fighting. However, decreasing the costs associated with conflict actually incentivizes states to fight more frequently. As such, the destructive power of nuclear weapons may ironically be better promoters of peace than even the most seasoned of diplomats.

3.4.4: Curveballs with a Runner on Third Base

In baseball, the duel between a pitcher and batter has the familiar matching pennies framework. The pitcher decides which type of pitch he wants to throw, while the batter anticipates a particular pitch type. If the batter guesses correctly, he is more likely to hit the ball. Thus, the batter wants a match, while the pitcher wants a mismatch.

Rather than rehash the matching pennies game for another time, let's instead consider a different strategic dynamic by putting a runner on third base. Suppose the pitcher on the mound only knew how to throw fastballs and curveballs. From the catcher's perspective, receiving a fastball is easier because the pitch travels in a straight line. Curveballs, as the name suggests, bend in midair and sometimes bounce before reaching home plate. If one of these curveballs trickles past the catcher, the runner on third can come home without the hitter lifting the bat off his shoulder. So throwing a curveball is inherently risky in this situation.

Thus, there are a couple of interesting questions here. As the pitcher's wildness increases, does the batter anticipate more fastballs? And does the pitcher, fearing that the runner on third will score, toss fewer curveballs? Think about these questions for a moment and make a couple of predictions.

To answer these questions, let's formalize a game from that setup. Similar to the penalty kicks game, suppose the batter earns 1 for a correct guess and 0 for an incorrect guess; the pitcher earns -1 if the batter guesses correctly and 0 if he guesses incorrectly. However, any time the pitcher throws a curveball, the batter gains x and the pitcher loses x, where x > 0. The value for x represents the probability the ball slips by the catcher on a curveball and the benefit of the run scoring to the batter.

The matrix looks like this:

	T Fast	T Curve
G Fast	1, -1	x, -x
G Curve	0, 0	1+x, -1-x

Let's first look for an instance of strict dominance. Note that x can take on any positive value and always harms player 2's payoff. Thus, many values of x make throwing a curveball a terrible strategy for player 2. We can see this if we isolate her payoffs, starting with her best response to the batter guessing a fastball:

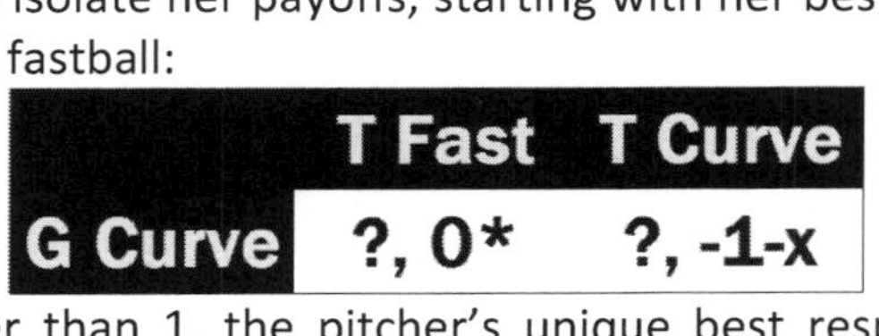

	T Fast	T Curve
G Curve	?, 0*	?, -1-x

If x is greater than 1, the pitcher's unique best response here is to throw a fastball since -x is greater than -1.

Now consider her best response to the batter guessing curveball:

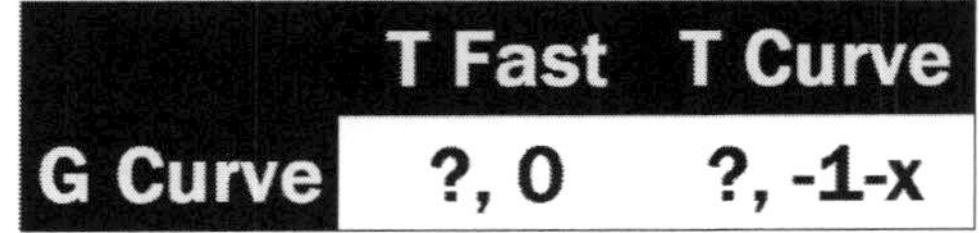

	T Fast	T Curve
G Curve	?, 0	?, -1-x

This time, regardless of the value of x, the pitcher's expected utility for throwing a curveball is negative. Meanwhile, her expected utility for throwing a fastball is 0. Thus, the precise value of x is irrelevant here: throwing a fastball is always the best response to the batter guessing curveball.

Combining these best responses together, we know that throwing a fastball strictly dominates throwing a curveball for $x > 1$. So when considering the universe of cases when $x > 1$, we can eliminate throwing curveball and focus on the reduced game:

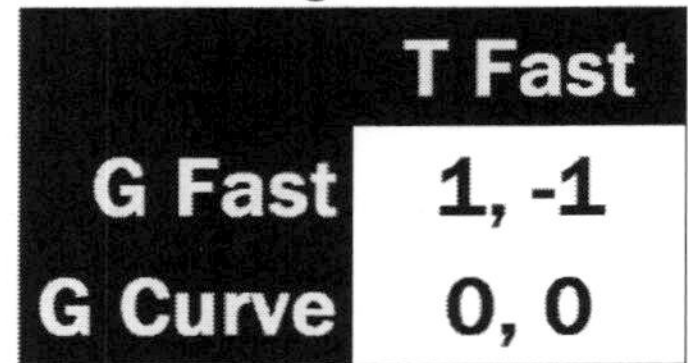

	T Fast
G Fast	1, -1
G Curve	0, 0

IESDS then instructs the batter to guess fastball. Therefore, the game has a unique equilibrium when $x > 1$: <guess fastball, throw fastball>.

Let's return to the pitcher's best response to the batter guessing a fastball:

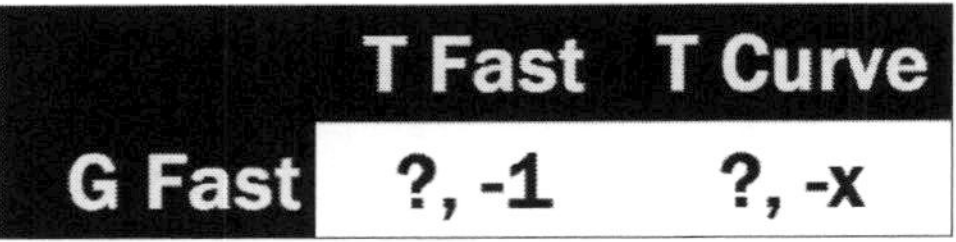

	T Fast	T Curve
G Fast	?, -1	?, -x

What about the cases when $0 < x \leq 1$? When x equals exactly 1, the pitcher is indifferent between throwing a fastball and throwing a curveball. Thus, $x = 1$ is a knife-edge condition. As Lesson 3.4 prescribed, we will ignore such a circumstance.

That leaves us with $0 < x < 1$ as the only case we have left to check. If we set x at such a value, the game's best responses look like this:

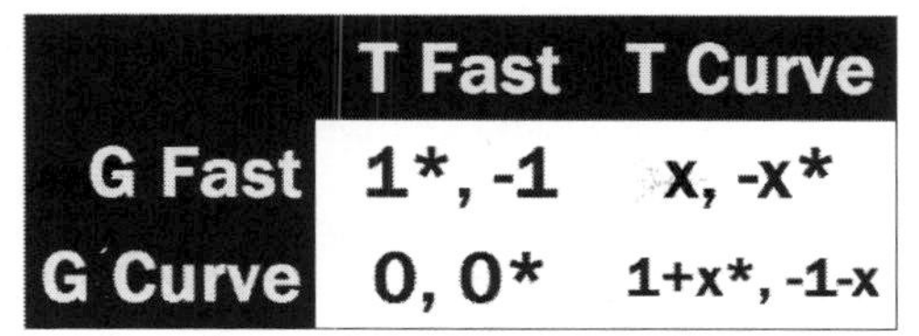

	T Fast	T Curve
G Fast	1*, -1	x, -x*
G Curve	0, 0*	1+x*, -1-x

The best responses show that the game has no pure strategy Nash equilibria. We also see the familiar matching pennies pattern of best responses, so we must turn to the mixed strategy algorithm to find the game's mixed strategy Nash equilibrium.

Let's start with the pitcher's expected utility for throwing a fastball:

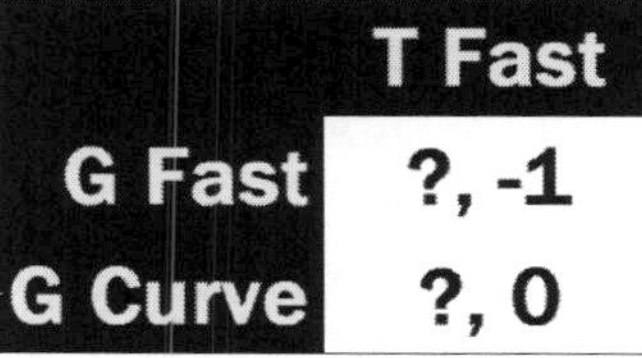

	T Fast
G Fast	?, -1
G Curve	?, 0

The pitcher earns -1 with probability σ_{GF} and 0 with probability $1 - \sigma_{GF}$. As an equation:

$$EU_{TF} = (\sigma_{GF})(-1) + (1 - \sigma_{GF})(0)$$

Now for the pitcher's expected utility for throwing a curveball:

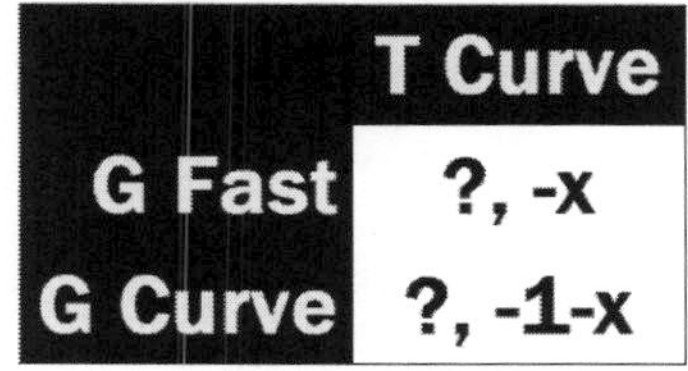

	T Curve
G Fast	?, -x
G Curve	?, -1-x

Here, she earns -x with probability σ_{GF} and $-1 - x$ with probability $1 - \sigma_{GF}$. As an equation:

$$EU_{TC} = (\sigma_{GF})(-x) + (1 - \sigma_{GF})(-1 - x)$$

As always, we set those expected utilities equal to each other and solve for the mixed strategy:

$$EU_{TF} = EU_{TC}$$
$$EU_{TF} = (\sigma_{GF})(-1) + (1 - \sigma_{GF})(0)$$
$$EU_{TC} = (\sigma_{GF})(-x) + (1 - \sigma_{GF})(-1 - x)$$
$$(\sigma_{GF})(-1) + (1 - \sigma_{GF})(0) = (\sigma_{GF})(-x) + (1 - \sigma_{GF})(-1 - x)$$
$$-\sigma_{GF} = -x\sigma_{GF} - 1 - x + \sigma_{GF} + x\sigma_{GF}$$
$$-\sigma_{GF} = -1 - x + \sigma_{GF}$$

$2\sigma_{GF} = 1 + x$
$\sigma_{GF} = (1 + x)/2$

Thus, the batter guesses fastball with probability (1 + x)/2 and guesses curveball with probability (1 - x)/2. These are valid probability distributions: the denominators are both positive, the numerators are both positive (since x is between 0 and 1), and the numerators are both less than the denominators.

So does the batter anticipate fastballs more frequently with a runner on third base? To find out, take the derivative of the batter's mixed strategy in equilibrium:

$\sigma_{GF} = (1 + x)/2$
$\sigma_{GF} = 1/2 + x/2$
$\sigma_{GF}' = (1/2 + x/2)'$
$\sigma_{GF}' = (1/2)' + (x/2)'$
$\sigma_{GF}' = (x/2)'$
$\sigma_{GF}' = 1/2$

The derivative of σ_{GF} is positive regardless of the value of x, so the probability the batter guesses fastball increases as x increases. In essence, the batter plays it safe. Even if he whiffs at the curveball, the runner on third will occasionally score. As such, he concentrates more frequently on the fastball.

While the batter's optimal strategy is intuitive, the pitcher's optimal strategy is comparatively strange. To solve for it, begin by finding the batter's expected utility for guessing fastball:

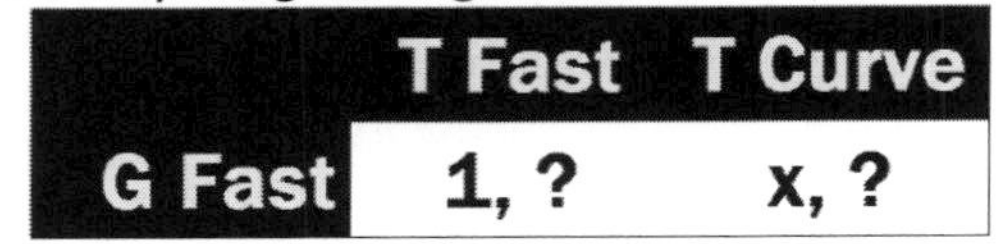

	T Fast	T Curve
G Fast	1, ?	x, ?

The batter earns 1 with probability σ_{TF} and x with probability $1 - \sigma_{TF}$. As an equation:

$EU_{GF} = (\sigma_{TF})(1) + (1 - \sigma_{TF})(x)$

Next, find the batter's expected utility for guessing curveball:

	T Fast	T Curve
G Curve	0, ?	1+x, ?

This time, he earns 0 with probability σ_{TF} and 1 + x with probability $(1 - \sigma_{TF})$. As an equation:

$$EU_{GC} = (\sigma_{TF})(0) + (1 - \sigma_{TF})(1 + x)$$

Now set those expected utilities equal to each other and solve for σ_{TF}:

$$EU_{GF} = EU_{GC}$$
$$EU_{GF} = (\sigma_{TF})(1) + (1 - \sigma_{TF})(x)$$
$$EU_{GC} = (\sigma_{TF})(0) + (1 - \sigma_{TF})(1 + x)$$
$$(\sigma_{TF})(1) + (1 - \sigma_{TF})(x) = (\sigma_{TF})(0) + (1 - \sigma_{TF})(1 + x)$$
$$\sigma_{TF} + x - x\sigma_{TF} = 1 - \sigma_{TF} + x - x\sigma_{TF}$$
$$\sigma_{TF} = 1 - \sigma_{TF}$$
$$2\sigma_{TF} = 1$$
$$\sigma_{TF} = 1/2$$

So when the pitcher throws fastballs half of the time and curveballs the other half of the time, the batter is indifferent between guessing fastball and guessing curveball. Notice that the x disappeared from this equation. Thus, the pitcher's optimal strategy does not vary on this interval! Even if she has great control of her curveball (say, x = .01) or is wild (x = .99), she still throws the curveball at the same frequency. But when her curveballs become too wild (x > 1), she suddenly stops throwing them altogether.

3.4.5: Comparative Statics of Take or Share (or Lack Thereof)

Not all games have interesting comparative statics. In fact, the equilibria of some games will not change at all even as you alter some of its features.

For example, suppose you were asked to discuss how the partially mixed strategy Nash equilibria of the take or share game change as a function of the value of the good v > 0. Recall the matrix of the game looked like this:

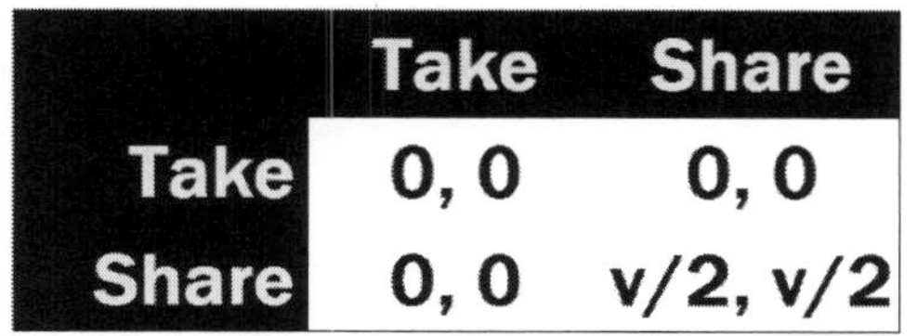

	Take	Share
Take	0, 0	0, 0
Share	0, 0	v/2, v/2

In the partially mixed strategy Nash equilibrium, one player selects take as a pure strategy while the other freely mixes between take and share. More explicitly, we can write those equilibria as <take, σ_{take}> and <σ_{take},

take>, where $0 < \sigma_{take} < 1$. But note that none of those equilibrium strategies change as a function on v. If a player plays σ_{take} for one value of v, he or she can play that exact same mixture for any other value of v. Thus, the exact value of v is trivial.

Do not over think these situations; such trivial comparative statics are well within the realm of possibility, and you need to be prepared to take a hands-off approach should they surface.

Takeaway Points

1) *Comparative statics* measure how changing a game's inputs alter its equilibrium outputs.
2) Some games lack interesting comparative statics.

Lesson 3.5: Generalizing Mixed Strategy Nash Equilibrium

Although we have been working with mixed strategies since Lesson 1.5, our applications have been limited. In fact, we have only found mixed strategy equilibria in which players mix between two strategies. But many games have three or more strategies. If we limited our knowledge of mixing to just the simple cases, we would be powerless whenever we encountered more complex games. Consequently, we must generalize our knowledge of mixed strategy Nash equilibrium.

This lesson covers two key features of MSNE. First, we will see that all pure strategies played in a mixed strategy must generate the same expected utility in equilibrium. Second, and as a result of the first fact, we can eliminate weakly dominated strategies from some potential equilibria without doing any messy mathematical work.

3.5.1: The Support of a Mixed Strategy

In games with a finite number of strategies, we say a pure strategy is in the *support of a mixed strategy* if and only if the probability of playing that pure strategy in the mixed strategy is positive.

To see what this means, consider the following game:

	Left	Center	Right
Up	1, -1	-1, 1	0, 0
Down	-1, 1	1, -1	5, 0

Player 2 has three strategies: left, center, and right. Suppose player 2 used the mixture σ_{left} = .1, σ_{center} = .3, and σ_{right} = .6. Then left, center, and right are all in the support of player 2's mixed strategy, since she plays each of them some percentage of the time. However, if player 2 switched her mixture to σ_{left} = .7, σ_{center} = .3, and σ_{right} = .0, right would no longer be in the support of her mixed strategy, as she would never select it.

In a general game, suppose the players mix in equilibrium. Then we immediately know something about the pure strategies in the support of the players' mixed strategies: they all yield the same expected utility in equilibrium.

Why is this the case? Recall that a Nash equilibrium is a set of strategies, one for each player, such that no player has incentive to change his or her strategy given what the other players are doing. Imagine player 1 mixed between up and down. In the above game, if up generated an expected utility of -.5 against player 2's mixed strategy and down generated an expected utility of .3, player 1 clearly would not want to mix. If he did mix, then whenever his mixed strategy told him to move down, he could move up instead. Such a deviation improves his payoff from some combination of -.5 to .3 to .3 with certainty. Playing up any portion of the time unnecessarily lowers his payoff. As a result, if player 1 mixes between two or more strategies, player 2's strategy must induce player 1 to be indifferent between those strategies.

We have been subtly using this principle since we started working with mixed strategy Nash equilibrium. Think back to the weighted matching pennies game, which we used to introduce the mixed strategy algorithm:

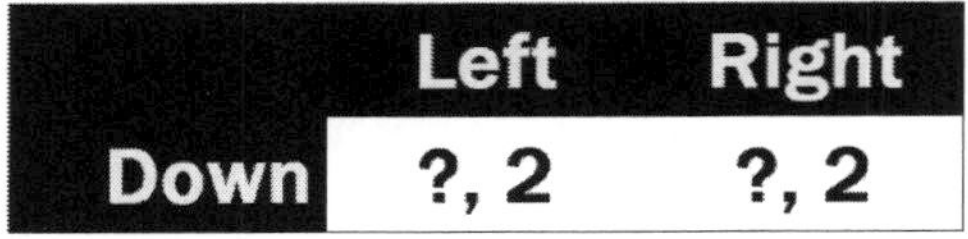

	Left	Right
Down	?, 2	?, 2

When we solved this game using the mixed strategy algorithm, we calculated each player's expected utility for his or her pure strategies as a function of the other player's mixed strategy. We then set those expected utilities equal to each other and solved for the mixed strategy that makes the equality possible. In essence, we found the mixed strategy that induces the other player's indifference.

The rule about indifference also prompted us to consider partially mixed strategy Nash equilibrium more carefully. Recall back to the matrix of Selten's game Lesson 1.8:

If player 1 selects down, player 2 always earns 2 whether she picks left or right:

	Left	Right
Down	?, 2	?, 2

Thus, by playing down as a pure strategy, player 1 has induced indifference in player 2's selection. Consequently, player 2 is free to mix, which is why we had to look at which mixtures still incentivize player 1 to maintain his down strategy.

3.5.2: A Necessary but not Sufficient Condition

It is necessary for all pure strategies in the support of a mixed to yield the same expected utility. However, it is *not* a sufficient condition. In other words, some other pure strategies may produce the same expected utility for a player but may not be played in a particular mixed strategy Nash equilibrium.

Consider the first game from last section, which is a revised edition of matching pennies:

	Left	Center	Right
Up	1, -1	-1, 1	0, 0
Down	-1, 1	1, -1	5, 0

If we ignore player 2's right strategy, the game is an exact duplicate of matching pennies. In the equilibrium of the original game, player 1 plays up with probability 1/2 and down with probability 1/2. Meanwhile, player 2 selects left with probability 1/2 and center with probability 1/2. Each player earns 0 in equilibrium.

Is this mixture still a Nash equilibrium of the revised game? Player 1 cannot profitably deviate; given player 2's mixed strategy, he earns 0 regardless of whether he chooses up or down. Since player 2 never selects right, player 1 cannot earn the 5 he receives if the players reach the <down, right> outcome.

Player 2 is similarly stuck. Given player 1's strategy, she earns 0 if she plays left and 0 if she plays center; the mixed strategy algorithm ensures this. She earns 0 if she plays right as well:

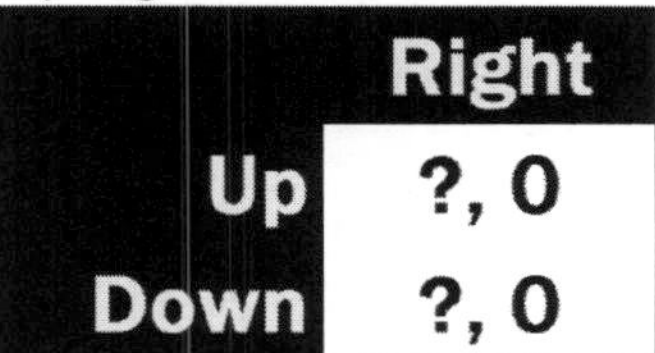

	Right
Up	?, 0
Down	?, 0

Half of the time, player 1 picks up and player 2 earns 0; the other half of the time, player 1 selects down and player 2 receives 0. Thus, player 2 cannot profitably deviate to right given player 1's mixed strategy. In turn, the equilibrium strategies of matching pennies are still a mutual best response to the revised game, so it is an equilibrium here as well.

We can now see why the rule about the strategies in the support of equilibrium mixed strategies is not a sufficient condition. If it were, player 2 would have to play right in her mixture, as it produces the same expected utility of 0 given player 1's mixed strategy. But the equilibrium we found has player 2 only mix between left and center.

3.5.3: A Trick with Weak Dominance

Weak dominance is fickle. If we eliminate a weakly dominated strategy from a game, any remaining Nash equilibria are also Nash equilibria in the unreduced game. However, some Nash equilibria may exist in the original game that are not in the reduced game. As a result, after eliminating weakly dominated strategies from the game, we still have to go back and solve the original game.

If we have to start over anyway, why waste effort finding weakly dominated strategies? As it turns out, knowing weakly dominated strategies makes finding mixed strategy equilibria substantially less time consuming. If a player mixes among *all* of his or her strategies, in a game with a finite number of strategies, the other player cannot play a weakly dominated strategy in equilibrium.

Why not? Remember back to the take or share game:

	Share	Take
Share	4, 4	0, 8
Take	8, 0	0, 0

If player 2 mixes between take and share, player 1 cannot choose share with positive probability. Whenever player 2 randomly selects take, player 1 earns 0 regardless of his choice:

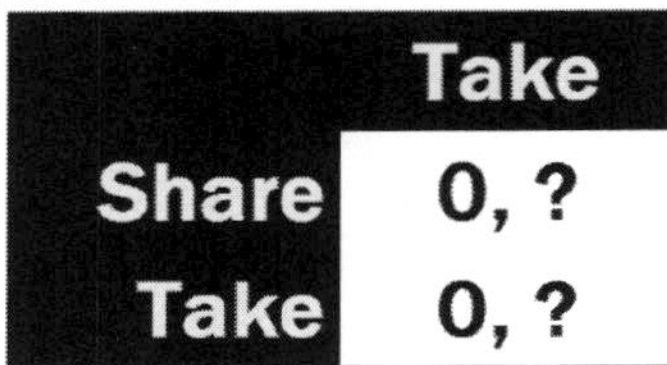

	Take
Share	0, ?
Take	0, ?

But when player 2 picks share, which occurs some percentage of the time, player 1 earns strictly more from choosing take:

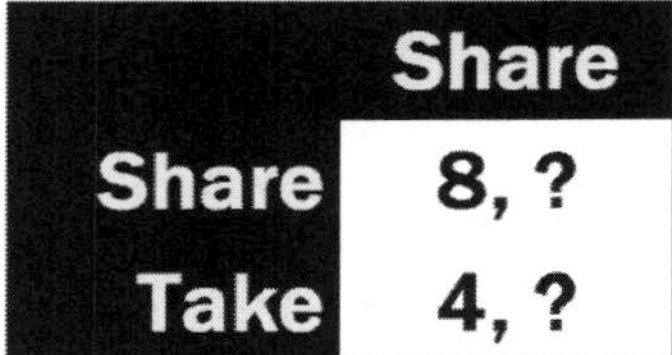

	Share
Share	8, ?
Take	4, ?

If player 1 knew that player 2 would always select take, he would be indifferent between share and take. But if player 2 is also playing share some percentage of the time, player 1 *must* play take. If he chooses share, he earns 4 some percentage of the time and 0 the rest of the time. But if he selects take instead, he earns 8 that same percentage of the time and 0 the rest of the time. In both of these cases, the "rest of the time" part cancels out, but the 8 from take is always greater than the 4 from share. The strategies in the support of a mixed strategy Nash equilibrium must yield the same expected utility. This is not possible if player 1 tries mixing as well; take provides a strictly greater amount than share if player 2 is mixing. As such, player 1 cannot mix if player 2 mixes.

The result generalizes beyond 2x2 games. Suppose strategy A weakly dominates strategy B for a particular player, and also suppose the opponent mixed among all n of his or her pure strategies. Let those strategies be 1, 2, ... , n which the mixing player selects with probabilities σ_1, σ_2, ... , σ_n. Additionally, let $EU_{A,1}$, $EU_{A,2}$, ... , $EU_{A,n}$ be the original player's expected utilities for when he or she plays A and his or her opponent plays the corresponding strategy. Then the original player's expected utility for A equals:

$$EU_A = (\sigma_1)[EU_{A,1}] + (\sigma_2)[EU_{A,2}] + \dots + (\sigma_n)[EU_{A,n}]$$

We can likewise define the original player's payoffs for B as follows:

$$EU_B = (\sigma_1)[EU_{B,1}] + (\sigma_2)[EU_{B,2}] + \dots + (\sigma_n)[EU_{B,n}]$$

By definition of weak dominance, $EU_{A,1} \geq EU_{B,1}$, $EU_{A,2} \geq EU_{B,2}$, and so forth. Since all of the probabilities σ_1, σ_2, ... , σ_n are greater than or equal to zero, it follows that $(\sigma_1)EU_{A,1} \geq (\sigma_1)EU_{B,1}$, $(\sigma_1)EU_{A,2} \geq (\sigma_1)EU_{B,2}$, and so forth. But at least one of these weak inequalities must hold strictly. Thus, when we sum all of them together, the expected utility for A must be greater than the expected utility for B. Consequently, the original player cannot mix between A and B; to be willing to do so, his or her expected utilities for A and B must be equal to one another, which we have shown is not the case.

This rule is extremely helpful when we break away from the 2x2 games. Consider this game, which we explored in Lesson 1.2:

	Left	Right
Up	0, 1	4, 2
Middle	0, 3	3, 3
Down	-2, 2	3, -1

Middle weakly dominates both up and down for player 1. Through iterated elimination of weakly dominated strategies, we discovered <middle, left> and <middle, right> were the only pure strategy Nash equilibria. But to fully solve the game, we have to search for partially mixed strategy and totally mixed strategy Nash equilibria.

Visually, we have to investigate four cases:

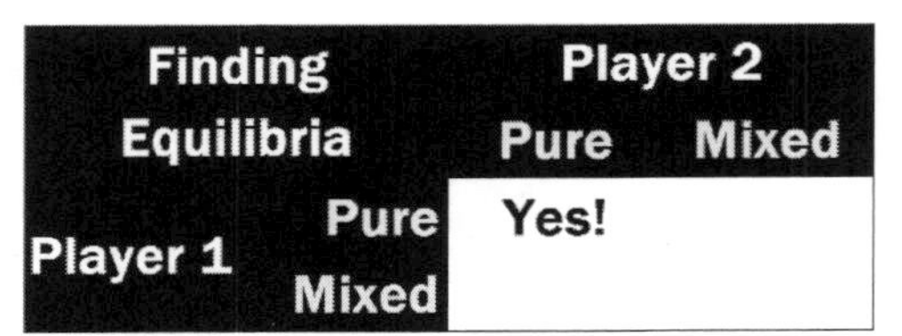

Finding Equilibria		Player 2 Pure	Player 2 Mixed
Player 1	Pure	Yes!	
	Mixed		

The top left category is pure strategy Nash equilibrium, which we have done already. The bottom right category is totally mixed strategy Nash equilibrium. Finally, the top-right and bottom-left categories are partially mixed strategy Nash equilibrium. Whenever we solve a game, we either must establish the existence of all equilibria or prove the non-existence of any equilibria in each of these categories.

We can rule out the totally mixed strategy case with ease. Recall that middle weakly dominates up and down; thus, middle is player 1's only strategy that is not weakly dominated. If player 2 is mixing, she must be playing all of her strategies. But if player 2 is playing all of her strategies and player 1 has weakly dominated strategies, player 1 cannot play those weakly dominated strategies. Therefore, the *only* strategy he can play is middle. But if we are interested in finding a totally mixed strategy Nash equilibrium, player 1 cannot play a pure strategy. As such, this game has no totally mixed strategy Nash equilibria:

Finding Equilibria		Player 2 Pure	Player 2 Mixed
Player 1	Pure	Yes!	
	Mixed		No!

Note that we could have shown no totally mixed strategy Nash equilibria exist by writing out the indifference requirements and solving for a mixed strategy that meets those requirements. However, that process would have been extremely time consuming. We would have had to consider *four* different types of mixtures from player 1: mixing among all three of his strategies, mixing only between up and middle, mixing only between up and down, and mixing only between middle and down. For once, weak dominance saved us time: knowing that up and down are weakly dominated meant we did not have to consider all of these cases.

Fortuitously, the process also left us close to solving for the partially mixed strategy Nash equilibrium in which player 1 plays a pure strategy and player 2 mixes. We know that if player 2 mixes, player 1 must play middle as a pure strategy. For player 2 to be willing to mix, she must earn the same expected utility for both pure strategies in the support of her mixture. In this case, she is indeed indifferent:

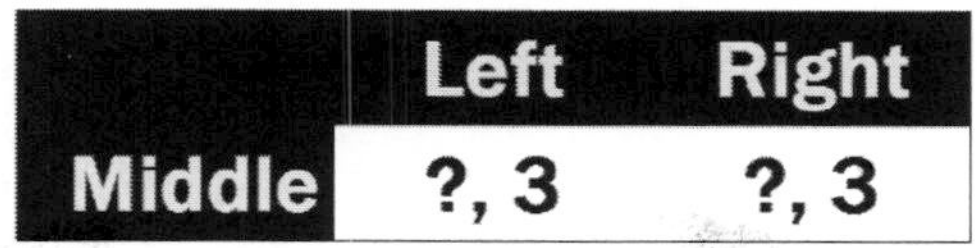

	Left	Right
Middle	?, 3	?, 3

Regardless of her strategy, player 2 always earns 3 if player 1 selects middle as a pure strategy. Thus, we have discovered a range of partially mixed strategy Nash equilibria in which player 1 selects middle as a pure strategy and player 2 mixes freely between left and right. That eliminates the top right box on our equilibria chart:

Finding Equilibria		Player 2 Pure	Player 2 Mixed
Player 1	Pure	Yes!	Yes!
	Mixed		No!

Now we must check for partially mixed strategy Nash equilibria in which player 1 mixes and player 2 uses a pure strategy. While the number of possible mixtures for player 1 is enormous, player 2 can only play left or right as a pure strategy. As such, let's consider the two cases based off of player 2's two pure strategies. First, suppose player 2 chose left:

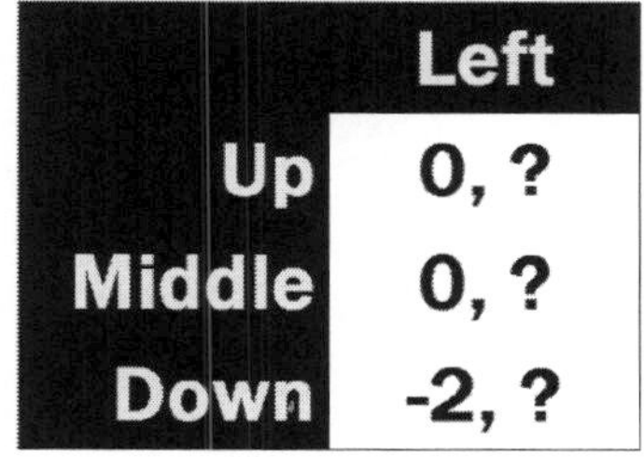

	Left
Up	0, ?
Middle	0, ?
Down	-2, ?

Recall that if player 1 is willing to mix, every pure strategy in the support of his mixed strategy must yield the same expected utility. But this implies player 1 cannot put positive probability on down, since up and middle provide a greater payoff. Thus, if player 2 selects left and player 1 is mixing, player 1's mixture must be between up and middle exclusively.

In turn, we need to find which mixtures between up and down (if any) induce player 2 to choose left as a pure strategy in equilibrium. We can find these mixed strategies by calculating player 2's expected utility for left and right as a function of player 1's mixture. First, let's find her expected utility for left:

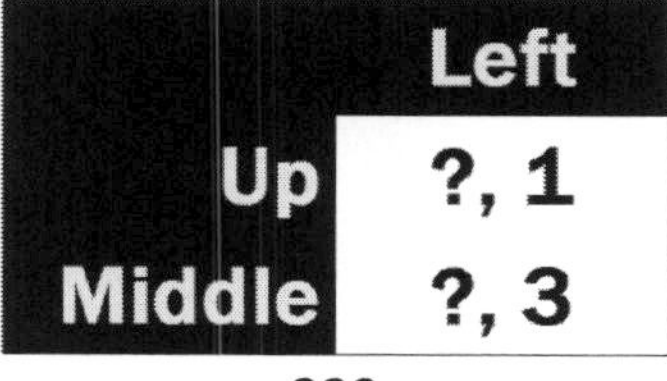

	Left
Up	?, 1
Middle	?, 3

Player 2 earns 1 with probability σ_{left} and 3 with probability $1 - \sigma_{left}$. As an equation:

$EU_{left} = (\sigma_{left})(1) + (1 - \sigma_{left})(3)$

Now suppose player 2 moved right:

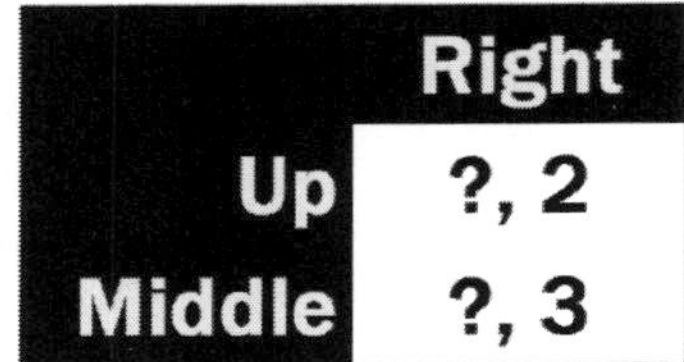

	Right
Up	?, 2
Middle	?, 3

This time, she earns 2 with probability σ_{left} and 3 with probability $1 - \sigma_{left}$. As an equation:

$EU_{right} = (\sigma_{left})(2) + (1 - \sigma_{left})(3)$

Player 2 is willing to play left as a pure strategy if her expected utility for left is at least as great as her expected utility for right:

$EU_{left} \geq EU_{right}$
$EU_{left} = (\sigma_{left})(1) + (1 - \sigma_{left})(3)$
$EU_{right} = (\sigma_{left})(2) + (1 - \sigma_{left})(3)$
$(\sigma_{left})(1) + (1 - \sigma_{left})(3) \geq (\sigma_{left})(2) + (1 - \sigma_{left})(3)$
$(\sigma_{left})(1) \geq (\sigma_{left})(2)$

Note that we are looking for a true mixture from player 1, so we know $\sigma_{left} > 0$. As such, we can divide by it:

$(\sigma_{left})(1) \geq (\sigma_{left})(2)$
$1 \geq 2$

We see a contradiction. Consequently, if player 1 is mixing between up and middle, player 2 is never willing to select left; she must always choose right. In turn, no partially mixed strategy Nash equilibrium exists in which player 2 selects left as a pure strategy.

The remaining case to check is whether player 2 can choose right as a pure strategy in a partially mixed strategy Nash equilibrium. Consider player 1's payoffs if she moves right:

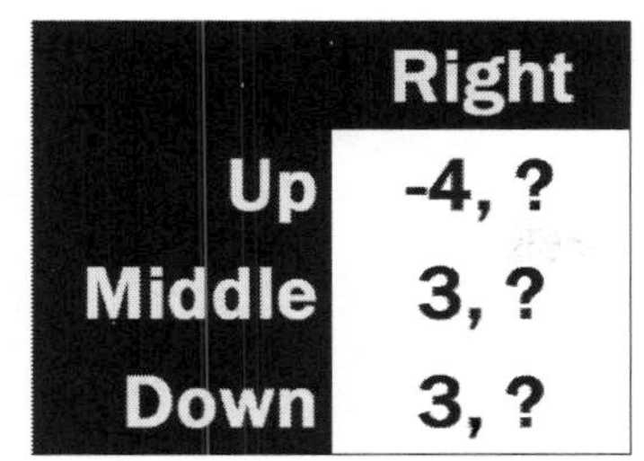

	Right
Up	-4, ?
Middle	3, ?
Down	3, ?

To be willing to mix, all pure strategies in the support of player 1's mixed strategy must provide the same expected utility. As a result, player 1 cannot play up if player 2 chooses right as a pure strategy. Instead, he must be mixing between middle and down.

Consequently, we must find which mixtures between middle and down (if any) induce player 2 to want to play right. Specifically, we must calculate player 2's expected utility for left and right as a function of player 1's mixed strategy and find which strategies make right provide at least as great of an expected utility as left.

Let's start with player 2's expected utility for right:

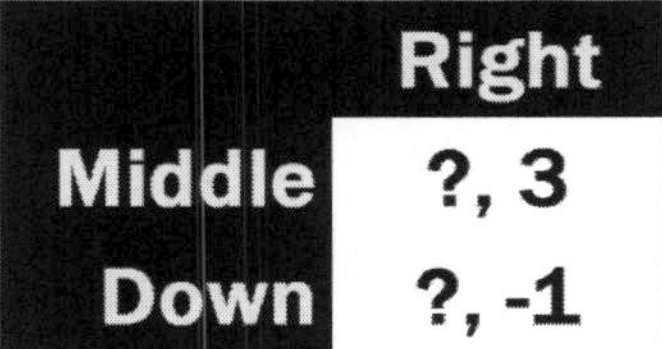

	Right
Middle	?, 3
Down	?, -1

Player 2 earns 3 with probability σ_{middle} and -1 with probability $1 - \sigma_{middle}$. As an equation:

$$EU_{right} = (\sigma_{middle})(3) + (1 - \sigma_{middle})(-1)$$

Now switch to player 2's expected utility for left:

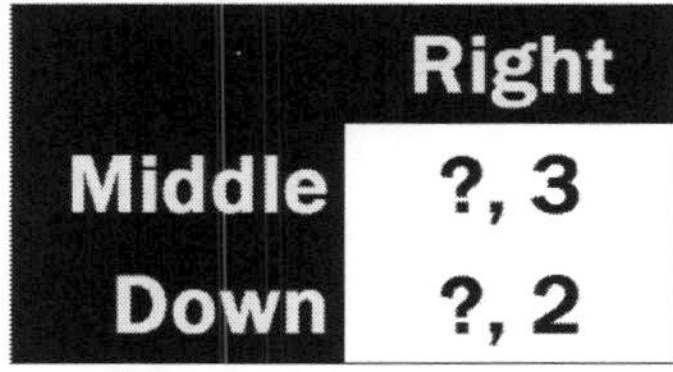

	Right
Middle	?, 3
Down	?, 2

This time, she earns 3 with probability σ_{middle} and 2 with probability $1 - \sigma_{middle}$. As an equation:

$$EU_{left} = (\sigma_{middle})(3) + (1 - \sigma_{middle})(2)$$

Player 2 is willing to play right if her expected utility for right is at least as great as her expected utility for left:

$EU_{right} \geq EU_{left}$
$EU_{right} = (\sigma_{middle})(3) + (1 - \sigma_{middle})(-1)$
$EU_{left} = (\sigma_{middle})(3) + (1 - \sigma_{middle})(2)$
$(\sigma_{middle})(3) + (1 - \sigma_{middle})(-1) \geq (\sigma_{middle})(3) + (1 - \sigma_{middle})(2)$
$(1 - \sigma_{middle})(-1) \geq (1 - \sigma_{middle})(2)$

Again, we are assuming player 1 is truly mixing, so $1 - \sigma_{middle} > 0$. Thus, we may divide by it:

$(1 - \sigma_{middle})(-1) \geq (1 - \sigma_{middle})(2)$
$-1 \geq 2$

We have derived a contradiction. -1 is less than 2, so player 2 is never willing to play right if player 1 mixes between middle and down. But that means player 2 using right as a pure strategy cannot be a part of a partially mixed strategy Nash equilibrium.

That rules out all possible cases of partially mixed strategy Nash equilibria in which player 2 selects a pure strategy. In turn, we can fill in the last portion of the equilibrium table:

Finding Equilibria		Player 2 Pure	Player 2 Mixed
Player 1	Pure	Yes!	Yes!
	Mixed	No!	No!

Thus, we are done with this game. Two PSNE exist, <middle, left> and <middle, right>, and there is a range of partially mixed strategy Nash equilibrium in which player 1 chooses middle as a pure strategy and player 2 mixes freely between left and right.

Takeaway Points

1) A pure strategy is in the *support of a mixed strategy* if the mixed strategy calls for playing that pure strategy with positive probability.
2) In a MSNE, all pure strategies in the support of a mixed strategy must yield the same payoff.
3) If the opposing player is mixing among all of his or her strategies, then the other player cannot play a weakly dominated strategy with positive probability in equilibrium.
4) Use a table to organize possible equilibrium configurations to keep your thoughts in order.

Lesson 3.6: Rock-Paper-Scissors

You have almost certainly played rock-paper-scissors (or roshambo) at some point in your life. In this lesson, we will formally derive its Nash equilibrium. Although you can likely guess that equilibrium, even slight changes to the payoffs quickly makes guessing the solution prohibitively difficult. From an empirical standpoint, these changes are important: a generalized version of rock-paper-scissors predominates the strategic setting of many video games and card games. Consequently, as the lesson progresses, we will work toward a general solution to rock-paper-scissors.

As always, it helps to draw the matrix of a game. Recall that rock smashes scissors, scissors cuts paper, and paper covers rock. A player holding the winning item receives 1 point while the loser earns -1. Players score 0 if they tie by revealing the same object.

If we fold together all of that information, we arrive at this extensive form:

	Rock	Paper	Scissors
Rock	0, 0	-1, 1	1, -1
Paper	1, -1	0, 0	-1, 1
Scissors	-1, 1	1, -1	0, 0

Let's mark the game's best responses. If your opponent is playing rock, your unique best response is paper; if your opponent is playing paper, your unique best response is scissors; and if your opponent is playing scissors, your best response is rock.

Using that intuition, we can place a star next to all of the winning payoffs to complete our goal:

	Rock	Paper	Scissors
Rock	0, 0	-1, 1*	1*, -1
Paper	1*, -1	0, 0	-1, 1*
Scissors	-1, 1*	1*, -1	0, 0

All of the 1s earn a star and everything else does not. However, no box contains a star for each payoff, which means no PSNE exists. As such, we must look into mixed strategies. Intuitively, you might guess that each player randomizes evenly among his or her strategies. That is, both player 1 and player 2 throw rock with probability 1/3, paper with 1/3, and scissors with 1/3.

Your intuition is correct. We can verify this by showing that such a probability distribution leaves both players indifferent among all three strategies. Let's look at player 1's expected utility for each of his pure strategies against player 2's mixed strategy, starting with rock:

	Rock	Paper	Scissors
Rock	0, ?	-1, ?	1, ?

If player 2 randomizes evenly among all three of her strategies, player 1 earns 0 with probability 1/3, -1 with probability 1/3, and 1 with probability 1/3. As an equation:

$EU_{rock} = (1/3)(0) + (1/3)(-1) + (1/3)(1)$
$EU_{rock} = -1/3 + 1/3$
$EU_{rock} = 0$

So player 1's expected utility equals 0. He earns the same for paper and scissors.

Here is the calculation for paper:

	Rock	Paper	Scissors
Paper	1, ?	0, ?	-1, ?

This time, he earns 1 with probability 1/3, 0 with probability 1/3, and -1 with probability 1/3. As an equation:

$EU_{paper} = (1/3)(1) + (1/3)(0) + (1/3)(-1)$
$EU_{paper} = 1/3 – 1/3$
$EU_{paper} = 0$

So his expected utility equals 0 once again.

Now consider scissors:

	Rock	Paper	Scissors
Scissors	-1, ?	1, ?	0, ?

Similar to before, he earns -1 with probability 1/3, 1 with probability 1/3, and 0 with probability 1/3. As an equation:

$EU_{scissors} = (1/3)(-1) + (1/3)(1) + (1/3)(0)$
$EU_{scissors} = -1/3 + 1/3$
$EU_{scissors} = 0$

Thus, we have shown player 1's expected utility equals 0 for each of his strategies. Thus, he may freely mix among them. If he chooses the same mixed strategy (1/3 rock, 1/3 paper, and 1/3 scissors), player 2 is similarly indifferent among her strategies, making it optimal to maintain her mixture as well. Consequently, the game reaches an equilibrium.

There are a couple of problems with this guessing method. So far, we know that a single mixed strategy Nash equilibria exists. We know nothing about the uniqueness of that equilibrium—another mixed strategy among all three strategies could also leave each player indifferent. Moreover, a player could possibly randomize between two strategies and ignore the third in equilibrium. To properly solve this game, we must have a formal way of deriving its equilibria.

To see the second issue, consider a slightly different version of the game:

	Rock	Paper	Scissors
Rock	0, 0	-2, 2	1, -1
Paper	2, -2	0, 0	-1, 1
Scissors	-1, 1	1, -1	0, 0

If we imagine the payoffs as dollars exchanged, this version of rock-paper-scissors now requires a rock player to fork over two dollars to a paper opponent.

The 1/3, 1/3, 1/3 mixture is no longer a Nash equilibrium; the change in expected utilities breaks the players' indifference. To see this, first consider player 1's expected utility for rock:

	Rock	Paper	Scissors
Rock	0, ?	-2, ?	1, ?

Now player 1 earns 0 with probability 1/3, -2 with probability 1/3, and 1 with probability 1/3. As an equation:

$EU_{rock} = (1/3)(0) + (1/3)(-2) + (1/3)(1)$
$EU_{rock} = -2/3 + 1/3$
$EU_{rock} = -1/3$

Next, consider player 1's expected utility for paper:

	Rock	Paper	Scissors
Paper	2, ?	0, ?	-1, ?

This time, he earns 2 with probability 1/3, 0 with probability 1/3, and -1 with probability 1/3. As an equation:

$EU_{paper} = (1/3)(2) + (1/3)(0) + (1/3)(-1)$
$EU_{paper} = 2/3 – 1/3$

$EU_{paper} = 1/3$

Already, we have a problem. Player 1's expected utility for paper is greater than his expected utility for rock. If a player is willing to mix, each strategy in the support of his mixed strategy must provide the same expected utility. Paper and rock generate different expected utilities, so they cannot be in the same mixed strategy.

The problem continues with scissors:

	Rock	Paper	Scissors
Scissors	-1, ?	1, ?	0, ?

These are the same payoffs as in the original version of rock-paper-scissors, so we know that player 1's expected utility is 0 here. That means paper provides the greatest expected utility against player 2's 1/3, 1/3, 1/3 mixture, which in turn implies that player 1 wants to choose paper as a pure strategy. But if player 1 uses paper exclusively, player 2's unique best response is to select scissors as a pure strategy. However, that leads to player 1 wanting to switch to rock exclusively. This begins a cycle. The mixed strategy Nash equilibrium from the previous version of the game simply does not work here.

We could try plugging in other mixed strategies, but the guess-and-check process is time consuming. Consequently, we need to develop a method which proves the uniqueness of equilibria and allows us to directly solve slightly different versions of rock-paper-scissors. The next two sections work toward that goal. First, we will learn about an important shortcut we can use with zero sum games which enables us to quickly rule out possible equilibrium strategies. Using that knowledge, we will then transition into a general form of rock-paper-scissors and derive its unique mixed strategy Nash equilibrium.

3.6.1: A Trick with Symmetric, Zero Sum Games

Rock-paper-scissors is a symmetric and zero sum game; each player has the same strategies and payoffs associated with those strategies, and the sum of both players' payoffs in each outcome is zero. Whenever a game meets these requirements, we know something important about the outcome of the game: *each player's expected utility must equal zero in equilibrium*.

Why is this true? Suppose it was not the case. That is, without loss of generality, suppose player 1's expected utility was positive and player 2's expected utility was negative in equilibrium. For our equilibrium conditions to hold, player 2 must not have a profitable deviation.

However, consider a strategy stealing argument. Player 2 could simply adopt player 1's strategy. Under this new strategy, the good outcomes occur equally as frequently as the bad outcomes for player 2. Thus, player 2's expected utility must equal zero. But if she previously earned a negative amount, this means that player 2 has a profitable deviation. Therefore, player 1 cannot have a greater expected utility than player 2 in equilibrium.

Since this argument works both ways, player 2 cannot have a greater expected utility than player 1 equilibrium either. The only way for both of these conditions to hold simultaneously is if each player earns 0 in equilibrium.

Armed with this knowledge, we can easily show that the players cannot mix between only two strategies in equilibrium in the basic rock-paper-scissors game. Let's focus on player 1's strategies, knowing that the symmetry of the game implies that these cases also cover player 2's strategies. We have three mixed strategies for player 1 to consider: (1) rock and paper but not scissors, (2) rock and scissors but not paper, and (3) paper and scissors but not rock.

Let's start by considering case 1:

	Rock	Paper	Scissors
Rock	?, 0	?, 1	?, -1
Paper	?, -1	?, 0	?, 1

Rather than deriving the precise best response, note that player 2 can choose paper and earn some positive amount. Let σ_{rock} be the probability player 1 chooses rock and $1 - \sigma_{rock}$ be the probability player 2 picks paper. Then player 2's expected utility for paper equals:

$$EU_{paper} = (\sigma_{rock})(1) + (1 - \sigma_{rock})(0)$$
$$EU_{paper} = \sigma_{rock}$$

Since player 1 is truly mixing, we know $\sigma_{rock} > 0$. Thus, whatever player 2's best response is, she must be earning a positive expected utility if player 1 only mixes between rock and paper. After all, paper earns a strictly positive amount. Any best response candidate that earns 0 or less must be worse than paper and is therefore not a best response.

Note that we are not claiming paper is definitely the best response. If we wanted to find the precise best response, we would have to do a little more digging. For example, if player 1 selects paper a vast majority of the time, scissors is a better response than paper for player 2. Although player 2 accepts some risk of losing by selecting scissors, she will win so much more frequently that it will make up for the difference. (Rock, of course, is strictly worse than paper against player 1's strategy, so it is out of the question.)

However, all of that information is superfluous once we know that player 2's best response to player 1's strategy generates a positive expected utility. Since the game is zero sum, player 1 must earn a negative amount when player 2 uses her best response. But because both players must have an expected utility of 0 in symmetric, zero sum games, player 1's original strategy cannot be a part of a mutual best response. Therefore, player 1 cannot use his mixture between rock and paper in a Nash equilibrium.

We can repeat this argument analogously with the other two cases. Let's check a mixture between rock and scissors:

	Rock	Paper	Scissors
Rock	?, 0	?, 1	?, -1
Scissors	?, 1	?, -1	?, 0

Now player 2 can select rock and guarantee a positive payoff. Although paper could possibly be a better response if player 1 relies more heavily on rock, that information remains superfluous. Regardless of player 2's exact optimal choice, player 1's expected utility must be negative. But this violates the rule about the equilibria of symmetric, zero sum games, so we can conclude that mixtures exclusively between rock and scissors will not work either.

Lastly, consider a mixture between paper and scissors:

	Rock	Paper	Scissors
Paper	?, -1	?, 0	?, 1
Scissors	?, 1	?, -1	?, 0

This time, scissors acts as the fail-safe. If player 2 uses it, she must earn a positive payoff. But just as before, that implies her best response generates a positive expected utility, which in turn gives player 1 a negative expected utility. Since player 1 cannot earn a negative expected utility in equilibrium, we can conclude that mixtures exclusively between paper and scissors will not work.

In the course of this lesson, we have shown that rock-paper-scissors has no pure strategy Nash equilibria or mixed strategy Nash equilibria in which a player only mixes between two strategies. We also correctly guessed the existence of a mixed strategy Nash equilibrium in which each player mixes among all three strategies. However, we have not proven the uniqueness of this equilibrium. The next section covers the uniqueness part by showing how to solve for a generalized version of rock-paper-scissors.

3.6.2: Generalized Rock-Paper-Scissors

Consider the following game:

	Rock	Paper	Scissors
Rock	0, 0	-x, x	y, -y
Paper	x, -x	0, 0	-z, z
Scissors	-y, y	z, -z	0, 0

To maintain the flavor of rock-paper-scissors, constrain x, y, and z such that x > 0, y > 0, and z > 0. This ensures that paper still beats rock, scissors still trumps paper, and rock still destroys scissors. However, by letting the exogenous variables be any strictly positive value, we can vary the lethality of each strategy against each other strategy. For example, if x is extremely large, then paper obliterates rock, and the rock player must hand a large sum of money to the paper player.

One interpretation of this game is like regular rock-paper-scissors, except different matchups of strategies results in different amounts of dollars exchanging hands. A more natural interpretation is of character selection in video games, particularly two dimensional fighting games. Characters have different strengths and weaknesses, which leads to good matchups against some opposing characters and bad matchups against others. Thus, a large value for x implies that the "paper" character has a strong matchup versus the "rock" character, and so forth.

Regardless of the interpretation, this game has no PSNE. We can see this by marking best responses:

	Rock	Paper	Scissors
Rock	0, 0	-x, x*	y*, -y
Paper	x*, -x	0, 0	-z, z*
Scissors	-y, y*	z*, -z	0, 0

Just like in the original version of rock-paper-scissors, all of the positive numbers receive an asterisk. However, no single outcome has two asterisks, so no pure strategy Nash equilibrium exists.

Using the method from the previous section, we can also show that no mixed strategy Nash equilibrium exists in which a player mixes exclusively between two strategies. To see this, we will consider player 1's strategies, though the symmetry of the game implies that everything below also applies to player 2.

Let's start with a mixture between rock and paper:

	Rock	Paper	Scissors
Rock	?, 0	?, x	?, -y
Paper	?, -x	?, 0	?, z

Paper ensures player 2 a positive expected utility, as x > 0. This implies that player 2's best response generates a positive expected utility. In turn, player 1 must earn a negative expected utility, so he cannot mix exclusively between rock and paper in equilibrium.

Now consider a mixture between rock and scissors:

	Rock	Paper	Scissors
Rock	?, 0	?, x	?, -y
Scissors	?, y	?, -z	?, 0

Here, rock does the heavy lifting, guaranteeing player 2 a positive expected utility because y > 0. That starts us down the now-familiar causal chain which eventually tells us that player 1 cannot use a mixture exclusively between rock and scissors in equilibrium.

Finally, suppose player 1 mixes between paper and scissors:

	Rock	Paper	Scissors
Paper	?, -x	?, 0	?, z
Scissors	?, y	?, -z	?, 0

This time, scissors is the fail-safe strategy, granting a positive expected utility because z > 0. As such, player 1 cannot mix exclusively between paper and scissors.

Rock-paper-scissors is a finite game, so it must have an equilibrium. Since said equilibrium is not in pure strategies or mixtures involving only two strategies, both players must be mixing among all three strategies. To solve for the mixed strategy that leaves the opposing player indifferent among all three of his or her strategies, we calculate a player's expected utility for all of his or her pure strategies, set those expected utilities all equal to one another, and solve for the critical mixture.

This will be an involved process, so let's take it one step at a time. First, we need to define player 2's mixed strategy. Let $\sigma_{rock} > 0$ be the probability she plays rock, $\sigma_{paper} > 0$ be the probability she plays paper, and $\sigma_{scissors} > 0$ be the probability she plays scissors. Since σ_{rock}, σ_{paper}, and $\sigma_{scissors}$ represent all the possible strategy choices, we know from Lesson 3.1 that $\sigma_{rock} + \sigma_{paper} + \sigma_{scissors} = 1$. Rather than continue to work with three unknown variables, we should make the substitution $\sigma_{scissors} = 1 - \sigma_{rock} - \sigma_{paper}$ and work with two instead. Thus, as we continue our analysis, player 2 plays rock with probability σ_{rock}, paper with probability σ_{paper}, and scissors with probability $1 - \sigma_{rock} - \sigma_{paper}$.

We can now write player 1's expected utilities for each of his three strategies. Let's begin with rock:

	Rock	Paper	Scissors
Rock	**0, ?**	**-x, ?**	**y, ?**

Player 1 earns 0 with probability σ_{rock}, -x with probability σ_{paper}, and y with probability $1 - \sigma_{rock} - \sigma_{paper}$. As an equation:

$$EU_{rock} = (\sigma_{rock})(0) + (\sigma_{paper})(-x) + (1 - \sigma_{rock} - \sigma_{paper})(y)$$

Next, consider his expected utility for paper:

	Rock	Paper	Scissors
Paper	**x, ?**	**0, ?**	**-z, ?**

This time, player 1 earns x with probability σ_{rock}, 0 with probability σ_{paper}, and -z with probability $1 - \sigma_{rock} - \sigma_{paper}$. As an equation:

$$EU_{paper} = (\sigma_{rock})(x) + (\sigma_{paper})(0) + (1 - \sigma_{rock} - \sigma_{paper})(-z)$$

Finally, suppose player 1 selects scissors:

	Rock	Paper	Scissors
Scissors	**-y, ?**	**z, ?**	**0, ?**

Now he earns -y with probability σ_{rock}, z with probability σ_{paper}, and 0 with probability $1 - \sigma_{rock} - \sigma_{paper}$. As an equation:

$$EU_{scissors} = (\sigma_{rock})(-y) + (\sigma_{paper})(z) + (1 - \sigma_{rock} - \sigma_{paper})(0)$$

We are searching for a combination of σ_{rock} and σ_{paper} that leaves player 1 indifferent among all three of his strategies. Thus, we set these expected utilities equal to one another and solve for σ_{rock} and σ_{paper}:

$$EU_{rock} = EU_{paper} = EU_{scissors}$$

$$EU_{rock} = (\sigma_{rock})(0) + (\sigma_{paper})(-x) + (1 - \sigma_{rock} - \sigma_{paper})(y)$$
$$EU_{rock} = -x\sigma_{paper} + y - y\sigma_{rock} - y\sigma_{paper}$$

$$EU_{paper} = (\sigma_{rock})(x) + (\sigma_{paper})(0) + (1 - \sigma_{rock} - \sigma_{paper})(-z)$$
$$EU_{paper} = x\sigma_{rock} - z + z\sigma_{rock} + z\sigma_{paper}$$

$EU_{scissors} = (\sigma_{rock})(-y) + (\sigma_{paper})(z) + (1 - \sigma_{rock} - \sigma_{paper})(0)$
$EU_{scissors} = -y\sigma_{rock} + z\sigma_{paper}$

Notice that the simplified expected utility function for scissors only has two terms. Moreover, its only term for σ_{rock} matches the only term for σ_{rock} in the expected utility for rock, and its only term for σ_{paper} matches the only term for σ_{paper} in the expected utility for paper. Therefore, we can link the expected utility of scissors to the expected utility of rock to solve for σ_{paper}, and we can link the expected utility of scissors to the expected utility of paper to solve for σ_{rock}. We will be finished afterward.

Although that may have sounded complicated, the actual math is fairly simple. Let's start by solving for σ_{paper}:

$EU_{rock} = EU_{paper} = EU_{scissors}$
$EU_{rock} = EU_{scissors}$
$EU_{scissors} = -y\sigma_{rock} + z\sigma_{paper}$
$EU_{rock} = -x\sigma_{paper} + y - y\sigma_{rock} - y\sigma_{paper}$
$-y\sigma_{rock} + z\sigma_{paper} = -x\sigma_{paper} + y - y\sigma_{rock} - y\sigma_{paper}$
$z\sigma_{paper} = -x\sigma_{paper} + y - y\sigma_{paper}$
$x\sigma_{paper} + y\sigma_{paper} + z\sigma_{paper} = y$
$\sigma_{paper}(x + y + z) = y$
$\sigma_{paper} = y/(x + y + z)$

Let's check the validity of this element of a probability distribution. x, y, and z are all positive values, so the numerator and denominator are all positive. The denominator is also greater than the numerator, as the inequality $x + y + z > y$ immediately reduces to $x + z > 0$. Thus, both of our rules hold, so σ_{paper} is a valid probability.

Now let's solve for σ_{rock}. Recall that the only element containing σ_{paper} in the expected utility for scissors matches the only element containing σ_{paper} in the expected utility for paper. Thus, we can easily eliminate σ_{paper} from the equations by setting the expected utilities for scissors and paper equal to each other:

$EU_{rock} = EU_{paper} = EU_{scissors}$
$EU_{scissors} = -y\sigma_{rock} + z\sigma_{paper}$
$EU_{paper} = x\sigma_{rock} - z + z\sigma_{rock} + z\sigma_{paper}$
$EU_{paper} = EU_{scissors}$
$-y\sigma_{rock} + z\sigma_{paper} = x\sigma_{rock} - z + z\sigma_{rock} + z\sigma_{paper}$
$-y\sigma_{rock} = x\sigma_{rock} - z + z\sigma_{rock}$

$x\sigma_{rock} + y\sigma_{rock} + z\sigma_{rock} = z$
$\sigma_{rock}(x + y + z) = z$
$\sigma_{rock} = z/(x + y + z)$

Again, we must check for the validity of this element of a probability distribution. As before, both terms are clearly positive. The denominator must be greater than the numerator as well, since $x + y + z > z$ reduces to $x + y > 0$. So σ_{rock} meets the validity requirements.

However, we are not finished checking validity. Every event in the probability distribution must be non-negative and they must all sum to 1. While our mixtures between two strategies trivially fulfilled these requirements, we can manually check the third element in the case of three strategies. Using our solutions for σ_{rock} and σ_{paper}, let's undo our substitution for $\sigma_{scissors}$:

$\sigma_{scissors} = 1 - \sigma_{rock} - \sigma_{paper}$
$\sigma_{rock} = z/(x + y + z)$
$\sigma_{paper} = y/(x + y + z)$
$\sigma_{scissors} = 1 - z/(x + y + z) - y/(x + y + z)$
$\sigma_{scissors} = (x + y + z)/(x + y + z) - z/(x + y + z) - y/(x + y + z)$
$\sigma_{scissors} = x/(x + y + z)$

For the same reasons σ_{rock} and σ_{paper} were valid elements of a probability distribution, $\sigma_{scissors}$ is as well: both x and $x + y + z$ are both positive, and $x + y + z$ is greater than x. Thus, all of the possible events in the probability distribution occur with a probability between 0 and 1.

Moreover, the probabilities of the events sum to 1. We achieved this through the construction of $\sigma_{scissors}$. Recall that $\sigma_{scissors}$ equals $1 - \sigma_{rock} - \sigma_{paper}$. Simple rearrangement yields $\sigma_{rock} + \sigma_{paper} + \sigma_{scissors} = 1$, which was the desired property.

Combined, these two facts allow us to finally conclude that we have found a mixed strategy that leaves the other player indifferent among all three pure strategies. Thus, if both players select rock with probability $z/(x + y + z)$, paper with probability $y/(x + y + z)$, and scissors with probability $x/(x + y + z)$, the game is in a mixed strategy Nash equilibrium.

In addition, this process proves the *uniqueness* of the equilibrium. As we solved for the MSNE, our indifference conditions produced unique values for σ_{rock} and σ_{paper}. If multiple mixtures could have accomplished this, the algorithm would have also generated those figures. We only received one value for σ_{rock} and σ_{paper} though, so we know the MSNE is the only

MSNE in which a player mixes among all three strategies. Since our previous work showed that no equilibria consist of pure strategies or mixtures between exactly two strategies, we know the aforementioned MSNE is the only MSNE.

As a final comment, it is interesting to note that the main determinant of each strategy's probability has nothing to do with that strategy. For example, recall that each player chooses scissors with probability $x/(x + y + z)$. The denominator is the same as the other probabilities; it is the sum of all of the benefits of victory. Intuitively, you might guess the x in the numerator has something to do with scissors, as the $x/(x + y + z)$ is each player's probability of selecting scissors in the equilibrium.

Of course, game theory has a nasty tendency of giving us unintuitive results. This is just the latest example. The value of x, in fact, has no direct connection to scissors! Take a second look at the game's matrix:

	Rock	Paper	Scissors
Rock	0, 0	-x, x	y, -y
Paper	x, -x	0, 0	-z, z
Scissors	-y, y	z, -z	0, 0

As the matrix shows, x is the benefit a paper player receives for beating rock. It does *not* appear in any of the payoffs involving a scissors player.

Nevertheless, it ends up being the biggest determinant of the equilibrium probability of scissors. Using our method of deriving comparative statics, note that the probability of playing scissors increases as a function of x:

$$\sigma_{scissors} = x/(x + y + z)$$
$$\sigma_{scissors}' = [x/(x + y + z)]'$$
$$\sigma_{scissors}' = [(x)'(x + y + z) - (x)(x + y + z)']/(x + y + z)^2$$
$$\sigma_{scissors}' = [(1)(x + y + z) - (x)(1)]/(x + y + z)^2$$
$$\sigma_{scissors}' = [(x + y + z) - (x)]/(x + y + z)^2$$
$$\sigma_{scissors}' = (y + z)/(x + y + z)^2$$

Of course, y + z is positive; x + y + z is also positive, and squaring it maintains this property. Thus, the derivative is always positive, which means the probability of playing scissors is increasing in x.

So why does x have the most effect on $\sigma_{scissors}$ when it does not directly affect any payoffs when a player chooses scissors? To reach a sensible answer, we must take a step back. The value for x represents paper's ability to smash rock. As x increases, a victorious paper player takes more from the losing rock player. Initially, you might think this would encourage the players to play paper more frequently and rock less frequently.

In practice, rational action requires the players to think one step further. All other things being equal, x does make paper more attractive. But the players can anticipate this. In turn, scissors becomes more viable as a way to counteract paper's strength against rock. In effect, the players balance out paper's advantage by increasing their frequency of scissors.

We can see this in the example game from earlier:

	Rock	Paper	Scissors
Rock	0, 0	-2, 2	1, -1
Paper	2, -2	0, 0	-1, 1
Scissors	-1, 1	1, -1	0, 0

If we match this game back to the generalized form, we have x = 2, y = 1, and z = 1. From the mixed strategy algorithm, we know that the players select rock with probability 1/4, paper with probability 1/4, and scissors with probability 1/2.

3.6.3: Mixed Strategies as Population Parameters

From an empirical standpoint, mixed strategies seem bizarre. Even with the advent of smartphones, people do not rely on randomizing devices to make their strategic decisions even if a Nash equilibrium tells them to. Naturally, game theorists spend a great deal of time addressing the discrepancy. Are people actually playing these games rationally if they are never randomizing?

Perhaps. We could interpret a mixed strategy Nash equilibrium as the population parameters of a larger game rather than a specific strategy of an individual in a two-player game. The video game interpretation of rock-paper-scissors provides an intuitive framework for this discussion. After all, it takes time to learn how to optimally play a particular character. The average gamer simply does not have the time to learn all of the nuances of the rock, paper, and scissors characters. Instead, he learns how to masterfully play a single character and effectively selects that character as a pure strategy.

Without loss of generality, let's call that pure strategy "rock." Under normal circumstances, playing rock as a pure strategy does not work in a Nash equilibrium. But the player's choice can still be rationally optimal. When the gamer logs into the online interface for his game, he joins thousands of players on the server. If an automated matchmaking system randomly picks his opponent, what his expected utility for the game?

Look back to the relevant portion of the matrix:

	Rock	Paper	Scissors
Rock	0, ?	-x, ?	y, ?

Let σ_{rock} be the portion of the population that plays rock, σ_{paper} be the portion of the population that plays paper, and $\sigma_{scissors}$ be the portion of the population that plays scissors. From last section's indifference equations, we know that the player's expected utility is 0 if $\sigma_{rock} = z/(x + y + z)$, $\sigma_{paper} = y/(x + y + z)$ and $\sigma_{scissors} = x/(x + y + z)$. But if those are the portions of other players using each strategy, the player also has an expected utility of 0 if he plays paper or scissors. Thus, his choice to play rock as a pure strategy is rational; he cannot choose a different strategy and achieve a greater expected utility.

Amazingly, we have an environment in which everyone can play a pure strategy yet no one has a profitable deviation. The trick, of course, is morphing the two player game into a game where nature randomly draws players from a large pool of individuals. The MSNE of the two player game informs us of the exact distribution necessary to maintain individual indifference in the game with a large number of players.

Takeaway Points

1) In symmetric, zero sum games, each player's payoff in equilibrium must equal 0.
2) Mixed strategies can be thought of as population parameters instead of single players randomizing over choices.

Lesson 4.1: Infinite Strategy Spaces, Second Price Auctions, and the Median Voter Theorem

Thus far, every game we have encountered has had two common features: (1) they had a finite number of pure strategies and (2) we could represent the games using a matrix or a game tree. Unfortunately, not all games are so simple. A solid introduction to game theory requires at least some discussion of games that fall outside of these categories, so we address these issues briefly in this chapter.

Before looking at specific examples and solving for equilibria, a short point of clarification is in order. We do *not* define games based on their matrices or trees. Rather, a game is merely a set of players, their strategies, an order of moves, and payoffs associated with those moves. We use only these illustrative representations because they make it easier to solve for equilibria.

However, to some degree, these matrices and trees are a crutch: we cannot easily create a matrix or a tree for all games. For example, recall the duopolistic game from Lesson 1.2, in which two firms had to their level of production:

	Zero	One	Two	Three	Four	Five
Zero	0, 0	0, 9	0, 14	0, 15	0, 12	0, 5
One	9, 0	7, 7	5, 10	3, 9	1, 4	-1, -5
Two	14, 0	10, 5	6, 6	2, 3	-2, -4	-2, -5
Three	15, 0	9, 3	3, 2	-3, -3	-3, -4	-3, -5
Four	12, 0	4, 1	-4, -2	-4, -3	-4, -4	-4, -5
Five	5, 0	-5, -1	-5, -2	-5, -3	-5, -4	-5, -5

We limited their options to producing 0 to 5 goods. Of course, that was a simplification—5 is the largest number that could conveniently fit within the margins. More accurately, a firm could produce an arbitrarily large number of goods. For this game, the simplification was harmless—producing more any more than 5 yields a negative payoff regardless of the other player's strategy. As such, 0 strictly dominates all values greater than 0, so (by iterated elimination of strictly dominated strategies) the unique equilibrium of the simplified game is also the only equilibrium of the expanded game.

Even so, we could imagine a market situation that would support 40, 100, or 12,000 goods produced without firms dipping into negative profits. At this point, we could not possibly draw out a payoff matrix in any reasonable amount of time. (We might run out of paper too!)

Consequently, one goal of this chapter is to learn how to solve for equilibria even if payoff matrices are out of the question.

4.1.1: A Simple Game

To start, consider the following game. Players 1 and 2 simultaneously select a whole number between 1 and 100. Each player's payoff is the product of those two numbers. What are the game's equilibria?

Before, our method would have required drawing out the game tree, starring best responses, and finding the mutual best responses. But notice that this process would take *forever*: each player has 100 strategies, so the game matrix contains 10,000 outcome boxes. Even if we finished writing the matrix, we would then have to mark best responses for 200 strategies. At least for this game, the game matrix has outlived its usefulness.

Fortunately, we can use the following algorithm to efficiently find this game's pure strategy Nash equilibria:

1) Consider a single pure strategy from one player.
2) Find the other player's best responses to that strategy.
3) Check whether the original strategy is a best response to any of those best responses. If so, those strategies are an equilibrium.
4) Repeat this process for *all* of the pure strategies from the original player.

Let's use player 2's strategies for the example. To start, suppose player 2 selected 1. Then player 1's payoff equals x, so he maximizes his payoff by selecting 100. But player 2's best response to 100 is not 1 but rather 100 as well. So the game has no PSNE in which player 2 selects 1.

What about player 2 picking 2? Then player 1's payoff equals 2x, so he maximizes his payoff by selecting 100 once again. But player 2's best response to 100 is still picking 100. So the game has no PSNE in which player 2 selects 2.

We could repeat this process for 3 to 99. Each time, player 1's best response is to choose 100. And each time, player 2's best response is to also select 100. As such, the game has no equilibria in which player 2 selects any value less than 100.

Now consider player 1's best response to 100. Like player 2, his payoff equals 100x. Since this payoff is also strictly increasing in the number he selects, he should select the largest number possible, or 100. This is a Nash equilibrium.

4.1.2: A Game with No Equilibria

Recall that Nash's theorem says that all finite games have at least one Nash equilibrium. So far, because we have dealt exclusively with finite games, every game we have seen has had at least one equilibrium. But now we will cross the border into the realm of infinite games, which may or may not have an equilibrium.

Consider a revised version of the game from the previous section. As before, both players select a number, and their payoffs are the product of those numbers. This time, however, they may select *any* number strictly greater than 0.

It takes little time to show that the game has no equilibria. Could a player play any value for x in equilibrium? No. Holding the other player's strategy as y, the original player could profitably deviate to x + 1 and receive strictly more. As such, x cannot be played in equilibrium. But x applies to *all* possible choices, so no equilibria exist.

Essentially, the problem here is that each player would like to keep inflating their strategy closer to infinity, since something slightly more than an arbitrarily large number is a greater payoff than an arbitrarily large payoff. This causes the notion of best responses to completely break down, ensuring no best responses exist. Without best responses, there cannot be mutual best responses, and in turn there cannot be Nash equilibria.

4.1.3: Hotelling's Game and the Median Voter Theorem

Although John Nash published his famous theorem in 1951, economists and mathematicians toyed around with similar concepts decades prior. One of the most famous early examples is Hotelling's game, named after Harold Hotelling, who published this model way back in 1929.

Hotelling's setup is as follows. Two vendors are selling identical ice cream on a beach for $2 per cone. The vendors own carts and must therefore choose where to set up shop. Since their products and prices are identical, the location is all that matters—beachgoers will purchase from whichever vendor is closest to them, and they will split the business evenly if the vendors are in the same location.

Suppose we label the westernmost point of the beach 0 and the easternmost point 1. Further, imagine that all of the beachgoers are evenly distributed along the interval. Where should the vendors place their carts?

Again, we encounter the problem of infinite strategy spaces. As a result, game matrices provide no relief here and we must instead use logical reasoning to find the game's equilibria.

Pinning down what must be true about equilibrium strategies sometimes drastically cuts the number of possible equilibrium strategy combinations. To begin, note that there is something special about the location 1/2. Imagine that the first owner set up shop at that location. No matter where the other vendor places his cart, the first vendor *must* receive half of the business. For example, suppose the other vendor set up anywhere to the left of 1/2:

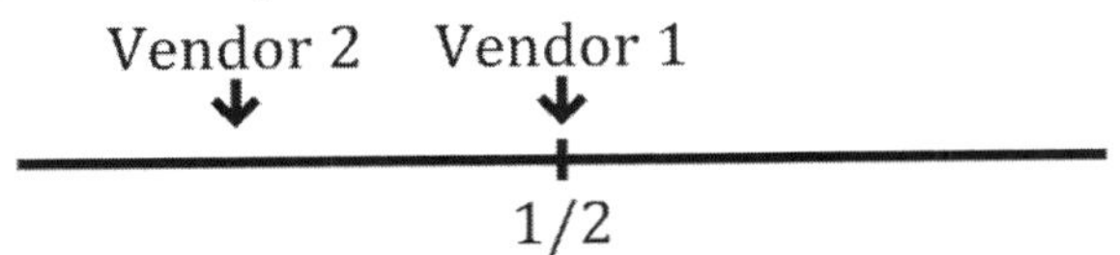

Vendor 1 receives *all* of the business from beachgoers on the right and a portion of the patrons on the left. Exactly how much of the business on the left he receives depends on precisely where vendor 2 locates herself. However, such specifics are irrelevant. All that matters here is that under such conditions, vendor 1 receives more than half the business.

Note that the same logic holds if vendor 2 places herself to the right of vendor 1. Now vendor 1 receives all of the left side business and a portion of the right side business, for a total of more than half.

Lastly, if vendor 2 also locates at 1/2, then the customers split evenly. As such, by placing himself at 1/2, vendor 1 guarantees himself half of the business.

Why does this tidbit matter? Recall that a Nash equilibrium is a set of mutual best responses. Thus, a set of strategies is *not* a Nash equilibrium if a player could profitably deviate to a different strategy.

In turn, consider any outcome in which vendor 1 does not receive at least half of the business. Then he could profitably deviate to positioning himself at 1/2, guaranteeing himself at least half of the revenue. As a result, in any Nash equilibrium, vendor 1 must receive at least half of the business.

Of course, the same logic applies to vendor 2—positioning herself at 1/2 also guarantees that at least half of the revenue will go to her. As such, she must also receive at least half of the business in any Nash equilibrium.

Keep in mind number of customers on the beach is not limitless—the sum of the vendors' business cannot exceed 1, since 1 is the total amount of space on the beach. So vendor 1 must receive at least half of the business, vendor 2 must receive half of the business, and the sum of the business must equal 1. This gives us three constraints:

$EU_1 \geq 1/2$

$EU_2 \geq 1/2$
$EU_1 + EU_2 = 1$

The only way to simultaneously satisfy all of these constraints is if both vendor 1 and vendor 2 receive *exactly* half of the business.

Take a step back for a moment. Originally, the game had a daunting number of possible strategies and a daunting number of possible payoffs for the vendors. Consequently, finding the Nash equilibria of the game seemed like an extremely difficult task. Yet the above constraints mean that we only have to consider cases where each vendor receives half of the customers. This is very manageable because this can only happen two ways: both choose a position equidistant from the halfway point or pick the same location.

Let's start with both equidistant from the halfway point. Since the game is symmetrical, let's say vendor 1 goes to the left and vendor 2 goes to the right. Those strategies look like this:

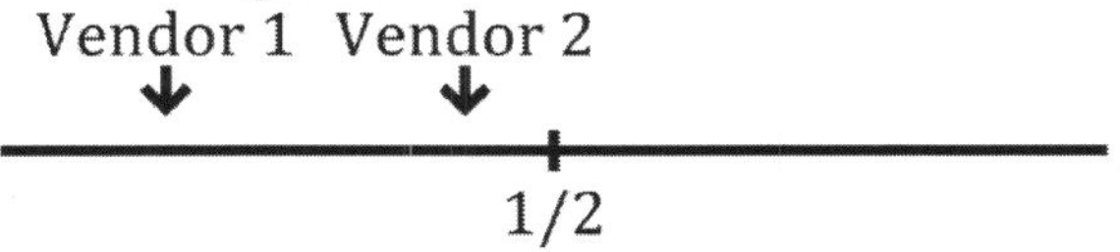

Although vendor 1 looks like he is at 1/4 and vendor 2 looks like she is at 3/4, the exact locations remain intentionally ambiguous. Indeed, we want to consider *all* such cases simultaneously. As such, the vendors could be at 0 and 1, 1/10 and 9/10, 1/3 and 2/3, or whatever.

In all of these cases, everyone to the left of 1/2 goes to vendor 1 and everyone to the right of 1/2 goes to vendor 2. For these strategies to form a Nash equilibrium, no vendor can have a profitable deviation.

But consider vendor 1's payoff if he moved his location to 1/2. He still receives all of the business on the left side, but now he steals some business away from the right side. This gives him more than half of the business, which is a profitable deviation. As such, the vendors cannot be equidistant from 1/2 in equilibrium.

This leaves both vendors selecting the same location as the only possibility. First, imagine they selected any point that is not 1/2. For example, imagine they were both to the left:

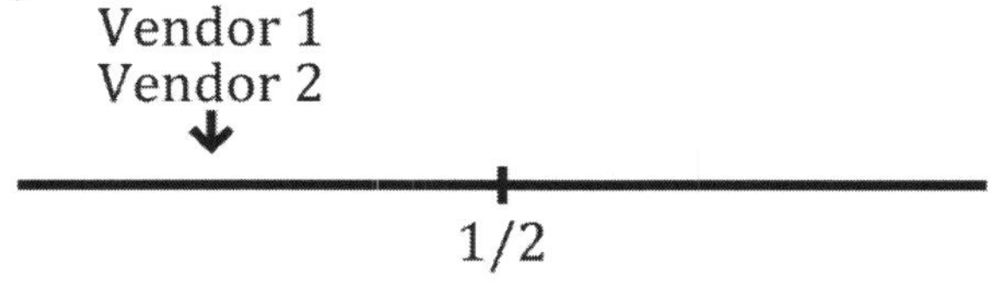

Again, they evenly split the business. However, vendor 1 could move himself to 1/2 instead. This secures all of the business on the right half as well as a portion of the business on the left. As such, these strategies are not an equilibrium.

The same is true if both vendors selected the same position to the right of 1/2. Now vendor 1 could move to 1/2 to obtain all of the business to the left and a portion of the business to the right.

This leaves just a single possibility: if both position themselves exactly at 1/2. Here, any deviation leaves a vendor with less than half of the business. Consequently, no profitable deviations exist. In turn, Hotelling's game has a unique equilibrium: the vendors occupy the same spot halfway along the beach.

This might be surprising, but we commonly see applications of Hotelling's game in everyday life. Consider, for example, gas station locations. It might seem strange that multiple gas stations frequently appear on the same intersection. Why don't they spread themselves out to make it easier for their customers? Hotelling's game provides the answer: despite being easier for customers, moving away from the center gives the gas station less business than it would by staying put. As such, gas station owners flock to the same location.

Similarly, consider Presidential elections in the United States. Immediately after the primary season ends, both the Democratic and Republican candidates dart toward the middle of the political spectrum. Political scientists note that this is also an application of Hotelling's game, known as the *median voter theorem*.

To win an election, candidates need half of the vote. The "median voter" represents the person whose political ideology places him with exactly half of the remaining voters to his left and exactly half of the remaining voters on his right. If a politician convinces the median voter to vote for him, then he guarantees himself half of the vote. If the rival takes a more liberal or conservative approach, he will certainly lose. As such, the other candidate must also mimic the median voter's preference to at least force a tie. Consequently, candidates race toward the center in the hopes of becoming unbeatable.

4.1.4: A Duel

Consider a duel between two gunslingers in the Wild West. Both have a gun with a single bullet and stand 100 yards apart. They slowly move toward each other until one fires a shot. The shooter survives if he hits.

Missing is lethal, however, as the other gunslinger can walk up and fire at point blank.

Naturally, we might wonder when the gunslingers will fire their shots. We might assume that more accurate shooters are willing to fire earlier while less accurate shooters will wait until close range. As it turns out, this theory is *wrong*—both gunslingers fire at the same time!

To see why, we need to formalize the problem. Players earn 1 if they survive and 0 if they die. Let $a_1(d_1)$ be the function that takes a **d**istance for player 1 and gives the **a**ccuracy of his shot from that distance. Likewise, $a_2(d_2)$ takes player 2's distance and gives her accuracy.

We only need to make two rather innocuous assumptions about those functions. First, suppose they are strictly decreasing; that is, the further the gunslinger is from the other one, the less accurate his shot will be. (Equivalently, the closer he is, the more accurate he is.) Second, let $a_1(0) = a_2(0) = 1$ and $a_1(100) = a_2(100) = 0$. That is, neither will hit from 100 yards but both are guaranteed to kill the other from point blank range.

Note that we are putting no other restrictions on these functions. The players could have identical skill-sets, one player could consistently be more accurate, player 1 could be better from short range but worse from long range, or player 2 could be better from short range but worse from long range. Surprisingly, we can solve for all of those cases in one motion.

We are ready to find the game's pure strategy Nash equilibria. Let $<d_1^*, d_2^*>$ represent those strategies, which are the distances the players shoot from if the other player has not already shot. It is easy to show that d_1^* must equal d_2^*. To see why, suppose not. Then either $d_1^* > d_2^*$ or $d_1^* < d_2^*$ is true. Without loss of generality, imagine it is $d_1^* > d_2^*$. Then player 1 shoots first, since his equilibrium strategy calls for him to fire at a distance further away than player 2's equilibrium strategy. Player 1 earns $a_1(d_1^*)$ in this alleged equilibrium.

However, consider a deviation to any distance between d_1^* and d_2^*. For the purposes of this illustration, consider in particular the midpoint between these two distances. Call that midpoint m. Suppose player 1 deviated to waiting to fire at m instead. Because $m > d_2^*$, player 1 still fires first. He receives $a_1(m)$ as his payoff. Since $a_1(d_1)$ is strictly decreasing and $m < d_1^*$, it must be true that $a_1(m) > a_1(d_1^*)$, meaning that this is a profitable deviation. In turn, $<d_1^*, d_2^*>$ cannot be an equilibrium.

If the math was not clear, this has an intuitive interpretation. Suppose you know that player 2 will wait until a particular distance to shoot. If you wanted to shoot first, would you ever want to shoot a yard before player 2's shooting distance? No, you would be better off waiting until a half yard before so you will be more accurate with your shot. But would you ever want to shoot from the half yard mark? No—again, you would be better off waiting a little bit longer. But this process repeats infinitely—as long as there is some distance before player 2 will shoot, even if that distance is arbitrarily small, you will still want to get slightly closer to improve your shot. So you can never optimally shoot your gun before your opponent does.

Consequently, any pure strategy Nash equilibrium requires $d_1^* = d_2^*$. Let's call that distance d*. What is the exact value for d*? Well, for player 1 to not want to deviate, he must be at least as well off from shooting at that moment as not shooting and hoping his opponent will miss, allowing him to walk up to her and fire at point blank. Note that the probability player 1 survives if player 2 shoots at d* equals $1 - a_2(d^*)$. As such, for player 1 to be willing to shoot at d*, it must be that:

$$a_1(d^*) \geq 1 - a_2(d^*)$$
$$a_1(d^*) + a_2(d^*) \geq 1$$

Player 1's other deviation is to fire before d*. As just mentioned, by waiting for player 2 to fire at d*, player 1 survives with probability $1 - a_2(d^*)$. For player 1 to not want to shoot earlier, it cannot be the case that his accuracy at d* is greater than his probability of survival at d*. After all, if it were greater, player 1 could shoot a split second earlier. His chance of hitting player 2 are greater than her chance of hitting him at d*, so player 1 is more likely to survive by deviating to the earlier time.

Formalizing this notion, it cannot be true that $a_1(d^*) > 1 - a_2(d^*)$. But this is the same thing as saying $a_1(d^*) \leq 1 - a_2(d^*)$ must be true. We can rework that inequality as follows:

$$a_1(d^*) \leq 1 - a_2(d^*)$$
$$a_1(d^*) + a_2(d^*) \leq 1$$

So the last two calculations tell us that $a_1(d^*) + a_2(d^*) \geq 1$ and $a_1(d^*) + a_2(d^*) \leq 1$ must simultaneously be true. The only way that can be the case is if $a_1(d^*) + a_2(d^*) = 1$.

The story is the same for player 2. For her to be willing to shoot at d*, she must be at least as well off from shooting at that moment as not shooting and hoping her opponent will miss, allowing her to walk up to him and fire at point blank. Note that the probability player 2 survives if player

1 shoots first at d* equals 1 – a_2(d*). Thus, for player 2 to be willing to shoot at d*, it must be that:

$$a_2(d^*) \geq 1 - a_1(d^*)$$
$$a_1(d^*) + a_2(d^*) \geq 1$$

Meanwhile, player 2 must also not want to fire before d*. For this to be true, it cannot be the case that her accuracy at d* is greater than her probability of survival at d*; otherwise, she would want to shoot slightly earlier and preempt player 1's shot. Formally, this means that $a_2(d^*) > 1 - a_1(d^*)$ cannot be true. But this is the same thing as saying $a_2(d^*) \leq 1 - a_1(d^*)$. We can manipulate this second inequality as follows:

$$a_2(d^*) \leq 1 - a_1(d^*)$$
$$a_1(d^*) + a_2(d^*) \leq 1$$

Notice that these are the same two inequalities we derived when checking for player 1's profitable deviations. The only way they can simultaneously hold is if $a_1(d^*) + a_2(d^*) = 1$. Thus, we have our answer. Both players fire at the same time, where that optimal time is the moment that the players' probabilities of killing the other sum to 1.

While Wild West duels may be a thing of the past, this game has a very relevant application to modern businesses. Many technologies meant to directly compete with each other (like PlayStation 4 and Xbox One or Blu-Ray and HD DVD) come out at roughly the same time. This is not a coincidence. Rather than probabilities of death and survival, imagine the duel recast as the probability a firm's technology captures the marketplace. Firms face a key tradeoff. The earlier you release your product, the more likely it is to have bugs. Yet releasing it later gives your rivals to the opportunity to establish market share.

But these are the exact same incentives the gunslingers faced. Therefore, we know the result: both firms release their products at the same time. Unfortunately for consumers, however, the incentives to preempt the other side's release date mean that both products have the potential to be bugged.

4.1.5: Cournot Duopolistic Competition

We now turn to a version of duopolistic competition from Antoine Augustin Cournot, a French economist who investigated how markets work when firms compete over quantities of goods produced. Unlike the model of duopolistic competition in Lesson 1.2, firms can choose *any* number of units to produce, including fractions of a whole number. As a result, we cannot use a payoff matrix to find equilibria.

Here is the game. Two firms simultaneously select a quantity of production. Call these values q_1 and q_2, with the subscript denoting the given firm's selection. Afterward, the market determines the price the firms can sell the good at. Let that price function be $900 - (q_1 + q_2)$. Notice that following the laws of supply and demand, the price decreases as the quantity increases. Firm 1's cost of production is 12 per unit and Firm 2's is 24. If each firm wants to maximize its profit, how much should each produce?

Unlike the duopolistic competition game from Lesson 1.2, this game has an infinite number of strategies, preventing us from using a payoff matrix to find the equilibria. But unlike Hotelling's game, each firm has a well-defined best response function. This allows us to use calculus to find the game's equilibrium. (Unfortunately, this also means you will not be able to do this problem if you do not know calculus.)

In fact, this type of game is common. A three-step algorithm produces the equilibrium:

1) Solve for each firm's utility function.
2) Convert each utility function to a best response function.
3) Use the best response functions as a system of equations and solve for the mutual best responses.

Steps 1 and 3 are straightforward for anyone who knows algebra. However, step 2 can be tricky because it requires multivariate calculus to solve the optimization problem.

First, we build each firm's utility function. Since the firms only wish to maximize profit, the formula is gross sales minus the firm's costs. For example, consider Firm 1's profits. The market price of the good is $900 - (q_1 + q_2)$, and it sells q_1 number of these goods. However it pays 12 to create each good. As such, its overall profit equals:

$EU_1(q_1, q_2)$ = [sale price][Firm 1's quantity] – [Firm 1's cost][Firm 1's quantity]

$EU_1(q_1, q_2) = [900 - (q_1 + q_2)](q_1) - 12q_1$

$EU_1(q_1, q_2) = (900 - q_1 - q_2)(q_1) - 12q_1$

$EU_1(q_1, q_2) = 900q_1 - q_1^2 - q_1q_2 - 12q_1$

$EU_1(q_1, q_2) = 888q_1 - q_1^2 - q_1q_2$

Similarly, Firm 2 sells q_2 of the good at price $900 - (q_1 + q_2)$ but subtracts 24 for each good it produces. Therefore, Firm 2's utility function equals:

$EU_2(q_1, q_2)$ = [sale price][Firm 2's quantity] – [Firm 2's cost][Firm 2's quantity]
$EU_2(q_1, q_2) = [900 - (q_1 + q_2)](q_2) - 24q_2$
$EU_2(q_1, q_2) = (900 - q_1 - q_2)(q_2) - 24q_2$
$EU_2(q_1, q_2) = 900q_2 - q_1q_2 - q_2^2 - 24q_2$
$EU_2(q_1, q_2) = 876q_2 - q_1q_2 - q_2^2$

This completes step 1.

Step 2 says to convert each utility function to a best response function. To understand why, notice that both firms' profits are a function of both their own production and their competition's. Consequently, formulating a good business plan means anticipating the other side's level of production and adopting the appropriate best response.

In the past, finding a best response simply meant looking along the corresponding row or column and finding the counter strategy that reaps the greatest rewards. Here, we only have an equation. But derivative calculus provides a solution: we can simply optimize the equation.

More mathematically, recall that Firm 1's profit function is $U_1(q_1, q_2) = 888q_1 - q_1^2 - q_1q_2$. Firm 1 can only control q_1; Firm 2 controls q_2. As such, Firm 1 must optimize q_1 to generate the greatest revenue given some value for q_2. But just requires taking the partial derivative with respect to q_1 and then optimizing the payoff function for q_1.

Here is the calculation for the partial derivative:

$(888q_1 - q_1^2 - q_1q_2)'$
$888 - 2q_1 - q_2$

To optimize, we set the partial derivative equal to 0 and solve for q_1:

$888 - 2q_1 - q_2 = 0$
$2q_1 = 888 - q_2$
$q_1 = 444 - q_2/2$

This is Firm 1's best response function. Put differently, it gives Firm 1's most profitable production quantity for any given production quantity of

Firm 2's. For instance, suppose Firm 2 produced 88 goods. Then Firm 1 receives the most profit by producing 444 – 88/2 units, or 400 units.

Step 2 of the algorithm also requires finding Firm 2's best response function, giving us an opportunity to practice the previous step a second time. Recall that Firm 2's profit function is $U_2(q_1, q_2) = 876q_2 - q_1q_2 - q_2^2$. Since Firm 2 only controls its own quantity produced, we begin by taking the partial derivative with respect to q_2:

$$(876q_2 - q_1q_2 - q_2^2)'$$
$$876 - q_1 - 2q_2$$

Then we optimize by setting the partial derivative equal to 0 and solving for q_2:

$$876 - q_1 - 2q_2 = 0$$
$$2q_2 = 876 - q_1$$
$$q_2 = 438 - q_1/2$$

This completes step 2.

The final step of the algorithm is to use the best response function as a system of equations and solve for equilibria. Note that the best response functions are two equations with two unknown variables. From basic algebra, this implies that the system of equations has a unique solution.

Why does this matter? Recall again that a Nash equilibrium is a set of mutual best responses. So imagine that q_1^* and q_2^* represented the equilibrium strategies. By definition, q_2^* must be a best response to q_1^*. In turn, substituting q_1^* into Firm 2's best response function *must* yield q_2^*. Likewise, q_1^* must be a best response to q_2^*, so substituting q_2^* into Firm 1's best response function *must* yield q_1^*. In other words, q_1^* and q_2^* *must* be the solution to the system of best response equations. Thus, we can find the game's unique equilibrium by finding the solution to the system of equations.

First, let's solve for q_1 using Firm 2's best response function:

$$q_2 = 438 - q_1/2$$
$$2q_2 = 876 - q_1$$
$$q_1 = 876 - 2q_2$$

Combining this with Firm 1's best response function, we can solve for q_2:

$q_1 = 444 - q_2/2$
$q_1 = 876 - 2q_2$
$444 - q_2/2 = 876 - 2q_2$
$3q_2/2 = 432$
$q_2 = 288$

So Firm 2 produces 288 in equilibrium.
We can then use $q_2 = 288$ to solve for q_1:

$q_2 = 288$
$q_1 = 876 - 2q_2$
$q_1 = 876 - 2(288)$
$q_1 = 876 - 576$
$q_1 = 300$

Thus, Firm 1 produces 300 in equilibrium, 12 more units than its competitor. Consequently, Firm 1's cheaper production costs grant it a larger market share than its competitor.

4.1.5: Second Price Auctions

We conclude with a brief look at auction theory. In a second price auction, individuals give sealed bids to the auctioneer. After, the auctioneer looks over all the bids and awards the good to the person with the highest bid. However, the winner pays an amount equal to the *second* highest bid.

For example, imagine Sotheby's auctioned off a copy of *Game Theory 101: The Complete Textbook*. Albert bid $10; Barbara bid $13; Charlie bid $0.13, and Danielle bid $30. Under a second price auction, Sotheby's would award Danielle the book, since her bid was the highest. But rather than paying $30 (as she would in a first price auction) she pays $13, the amount the second highest bidder submitted.

Second price auctions have a large number of Nash equilibria. However, we focus on one in particular: when everyone submits the maximum price they are willing to pay for the good.

Verifying that this is an equilibrium takes only a couple steps. First, consider the losers' strategies. Reducing their bids does not change their welfare—they still lose and do not pay anything. Increasing their bids may or may not change their welfare. If a loser increases his bid to a still losing price, he still loses and receives nothing. On the other hand, a loser increasing his bid to a winning price gives him the good—but at a price higher than the maximum he was willing to pay. As such, losers have no incentive to change their bids.

Now consider the winner's strategy. Increasing his bid does not change anything, as he will still win and continue paying the price of the second bid. Meanwhile, decreasing his bid can only hurt him. If he stays above the second highest bid, he still wins and still pays the second highest bidder's price. But if he drops below the second highest bid, he now loses the auction he otherwise would have won at an acceptable price. Consequently, the winner cannot profitably deviate either.

This equilibrium is remarkable for a number of reasons. First, it is strategy-proof. Just about every game we have covered in this textbook requires active thought from the players—they must think about what others are doing, strategize in response, and then worry about how competitors will strategize in response to strategizing in response. Players in a second price auction can be comparatively oblivious. Because submitting their maximum prices is weakly dominant, bidders will never regret having told the truth.

Second, submitting one's maximum weakly dominates all other strategies. This follows directly from the fact that a winner cannot control the price he pays (since the second highest bidder determines the payment) and that a loser receives nothing and cannot win the good without spending more than his value for it.

Third, it is honest. The bidders simply tell the auctioneer their true value for good.

Fourth, number of bidders does not matter. Whether there are two or two million, submitting the maximum price remains optimal.

Lastly, bidders need not know others' maximum prices. This is an important final note for this textbook. Throughout, we have assumed players have *complete information*—that is, they know each others' payoffs, they know they know each other's payoffs, and so forth. While complete information can go a long way, many interesting interactions involve one or both sides being in the dark. In an auction setting, for instance, I might not know how much you are willing to spend, and you might not know how much I am willing to spend. Sometimes incomplete information can drastically change the outcome of an interaction. However, for a second price auction, it does not—everyone can still safely submit their maximum price.

Takeaway Points

1) Matrices and game tree do not define games; players, strategies, timing, and payoffs do.
2) Infinite games cannot be easily drawn and therefore require other algorithms to solve.
3) Some infinite games have no equilibria.

Final Thought

This book explained all of the tools you will need to solve introductory level questions about strategic and extensive form games. Good luck in applying them.

And always remember: weak dominance will be the bane of your existence.

About the Author

William Spaniel is a PhD candidate in political science at the University of Rochester, creator of the popular YouTube series *Game Theory 101*, and founder of gametheory101.com. You can email him at williamspaniel@gmail.com or follow him on Twitter @gametheory101.

Made in the USA
San Bernardino, CA
02 October 2014